This book is the official history of the Anglo-Australian Telescope in New South Wales which came into operation in 1974. The telescope is part of the Anglo-Australian Observatory which provides facilities for research in optical astronomy for scientists from Britain and Australia.

T0281738

The Creation of the
Anglo-Australian Observatory

THE CREATION OF THE ANGLO-AUSTRALIAN OBSERVATORY

S.C.B. Gascoigne
K.M. Proust
M.O. Robins

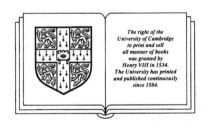

The right of the
University of Cambridge
to print and sell
all manner of books
was granted by
Henry VIII in 1534.
The University has printed
and published continuously
since 1584.

CAMBRIDGE UNIVERSITY PRESS
Cambridge
New York   Port Chester
Melbourne   Sydney

CAMBRIDGE UNIVERSITY PRESS
Cambridge, New York, Melbourne, Madrid, Cape Town, Singapore, São Paulo

Cambridge University Press
The Edinburgh Building, Cambridge CB2 2RU, UK

Published in the United States of America by Cambridge University Press, New York

www.cambridge.org
Information on this title: www.cambridge.org/9780521353960

© Cambridge University Press 1990

First published 1990
This digitally printed first paperback version 2005

A catalogue record for this publication is available from the British Library

ISBN-13 978-0-521-35396-0 hardback
ISBN-10 0-521-35396-3 hardback

ISBN-13 978-0-521-02019-0 paperback
ISBN-10 0-521-02019-0 paperback

The original text was computer typeset at the Anglo-Australian Observatory
by Katrina Proust.

Contents

Foreword

During the post-war years, some major astronomical developments occurred in Australia and the United Kingdom. In particular, radio astronomy groups were formed in each country which were widely accepted as being among the best in the world. The exploitation of the radio window in the earth's atmosphere initiated a revolution in astronomy which has continued with the advent of space research, opening up the whole of the electromagnetic spectrum to astronomical observation. It was soon apparent that the insights provided by the new data could only be realised if complementary optical observations of the finest quality were also available. Nowhere was this lesson brought home more to the British and Australian communities than in the example of the discovery of the first quasar 3C 273 in the early 1960s, a milestone in the development of astronomy. This radio source was observed and catalogued in the third Cambridge sky survey, and the Jodrell Bank group showed it was of extremely small angular size. Then, using lunar occultations, its exact position was measured with the Parkes radio telescope, thereby allowing its identification with a thirteenth magnitude blue 'star'. However, neither community had access to an optical telescope which was large enough or sensitive enough to obtain a spectrum of the object. This privilege passed to the Palomar Telescope and the resulting spectrum revealed the highest redshift ever observed, establishing 3C 273 to be the most distant and most luminous object in the Universe known at that time.

Against this background, it is not surprising that a heavy groundswell developed in each country in favour of building a large, modern optical telescope, a project which had the full support of radio, space and theoretical astronomers, as well as their optical colleagues. A number of options were considered which ultimately converged on an Australian–United Kingdom agreement to build a 3.9-metre telescope on Siding Spring Mountain in New South Wales, thereby giving the additional advantage of being able to study the relatively unexplored southern sky. That telescope, named the Anglo-Australian Telescope, was inaugurated in 1974 by the Prince of Wales. Fifteen years later, this book outlines the events leading up to that ceremony and the achievements since. By any standards, the AAT has been and continues to be an outstanding success; it can lay claim to being the best instrumented telescope in the world with a very wide capability and high sensitivity. The efforts and vision of those people who played the key roles during the 1960s and 1970s have been fully justified.

In parallel with the construction of the AAT, the UK Science and Engineering Research Council built a Schmidt Telescope and installed it nearby on the same mountain. This wide-field telescope was fully complementary to the AAT and their co-location led to many highly productive collaborative research programmes; for example, they have proved to be the most successful combination of telescopes in the world for finding large samples of quasars and for the discovery of the most distant objects in the Universe; the most recent record is a redshift of 4.42 – a far cry from those early days and 3C 273! The close association of the AAT and ST has been made even closer by their merger in 1988 to make the Anglo-Australian Observatory a two-telescope observatory, an event described in this history.

As the book reveals, the AAT had to face many problems and difficulties, not only those which always beset any major new project but additional ones posed by a bi-national venture involving two governments located on opposite sides of the world. The early days saw considerable controversy and difference of opinion, sometimes between the two partners, sometimes between institutions, and sometimes between the philosophies of individuals. It is to the credit of all involved that these problems were fully resolved to leave us today with a major front rank observatory managed with great harmony and good will on behalf of the two astronomy communities. We can attest to this from first-hand experience as a past Chairman and the present Chairman, a responsibility we both consider to be among the most rewarding and pleasant we have held.

J. Paul Wild, AC, CBE, FRS   Sir Robert Wilson, CBE, FRS
Canberra   University College London

Preface

Compared with many astronomical institutions in the northern hemisphere, the Anglo-Australian Observatory is young, but the story of the Observatory goes back well before regular scheduled observing commenced on the Anglo-Australian Telescope in June 1975. This is a story which begins in the 1950s when scientists in both Britain and Australia were considering construction of a large optical telescope in the southern hemisphere along the lines of the Palomar 200-inch Telescope. Our story follows the discussions and problems which arose during the 1960s among prospective partners, and which led at the end of the decade to the Anglo-Australian Telescope Agreement. We continue through the Observatory's first 14 years until it became a two-telescope observatory with the AAT and Schmidt Telescope.

Any project the size and the cost of the AAT obviously has to negotiate difficult paths. Indeed the AAT, which cost almost $16 million to build and took seven years to complete, presented new and challenging problems of engineering design and construction. Its outstanding scientific and technical team solved these problems, and today the AAT is celebrated for the perfection of its optics, the precision of its mounting, and the power and flexibility of its computer control system, in all of which it broke new ground. These are well known and published technical features.

At the time of commissioning in 1974, the AAT was famous throughout the astronomical world for a long and bitter dispute over how it would be run. This presented problems to rival those with which the engineers were grappling. The details of this debate are not so well-known, and previously have not been published in full. The Board's bi-national character presented an additional dimension to the problem.

This book covers both the technical and scientific, and the political and administrative history of the Anglo-Australian Observatory without which its story would be incomplete. This has meant that the authors have applied themselves more or less to certain areas. Gascoigne has written all the technical and scientific history and the technical matter of the AAT. Proust (in Australia) and Robins (in England) have supplied the non-technical content from the point of view of their respective countries.

Our sources have been many and varied, but have included official records of the Government agencies in each country which have been responsible for the initial and continued financing of the telescope. We have consulted the quarterly reports and other technical documents of the Project Office which take up some 21 shelf-metres of space in Australian Archives; minutes of meetings of the Joint Policy Committee and of the Board, the Board's Annual Reports and other records; the personal files of Gascoigne, Wehner and Eggen; and the AAT files and personal correspondence of Redman found in the archives of the Royal Greenwich Observatory. Sir Fred Hoyle's *The Anglo-Australian Telescope* was valuable both as a factual record and for the insight it gives into many of the deliberations of the Board. All this adds up to an enormous amount of information, and it has been impossible to describe all matters or to mention every person who could have been included in the AAT's history.

However, many people have given us their time and shared reflections of their association with the AAT project, and these have been invaluable in producing this book. We were especially impressed by the very large number of people who encouraged us to write the *whole* story, including the less pleasant side of the negotiations with the Australian National University in the early and difficult days. To them all we express our thanks indeed.

The Anglo-Australian Telescope Board supported the writing of this book as a project inititated by its Secretary and inspired by Taffy Bowen in 1983. Taffy Bowen urged that the AAT history was one that must be written before the unrecorded, personal memories of those who played an important part in it were lost. The Secretary especially thanks the Board for its support and confidence that she could bring off the project, and David Malin at the Anglo-Australian Observatory who was always there with help and encouragement.

We thank past and present Board members for their personal reflections and advice, particularly J.F. Hosie, Taffy Bowen, Sir Fred Hoyle, Paul Wild, R.J. Street, V.C. Reddish, Olin Eggen, K.N. Jones, Malcolm Longair, Sir Robert Wilson, Harry Atkinson, Greg Tegart, John Carver and Sir Francis Graham-Smith.

At the Anglo-Australian Observatory, David Malin and David Allen gave meticulous and constructive criticism of the manuscript, and Peter Gillingham and Doug Cunliffe were always available to provide some valuable background detail based on their long associations with the Observatory. Russell Cannon helped with the early history of the Schmidt Telescope at Siding Spring arising from his personal involvement in that project, and his door was always open for discussion of many parts of the text. John Straede and Bob Dean provided much useful advice on the section about the AAT's computer systems along with Patrick Wallace in Britain, and Maston Beard, formerly of CSIRO in Sydney. André Porteners produced most of the line drawings for the technical chapters.

Both Herman Wehner, at Mount Stromlo Observatory, and Harry Minnett, formerly of CSIRO's Division of Radiophysics, spent many hours with us discussing complex aspects of the AAT's engineering. We record our appreciation to the late Don Collins, former Site Engineer of Macdonald Wagner and Priddle at Siding Spring. Before his death in 1989 he gave the Observatory his large and splendid collection of photographs covering the complete construction phase of the telescope. Thanks go to Jack Rothwell, a member of the AAT Project Office, for his time discussing the telescope's engineering.

We acknowledge also the help received from Donald Morton, AAO Director for almost ten years, and Colin Plowman, Assistant Vice-Chancellor at the Australian National University, as well as from Brian Robinson, R. Hanbury Brown, John Whiteoak, Dick McGee, John Dawe and Louise Turtle. We are most grateful for assistance from Doreen Goodsell and Margaret Brown who gave us some very useful papers belonging to their husbands, after Bill Goodsell and David Brown died.

In governmental circles, the then Australian Department of Science, and the British Science and Engineering Research Council gave us access to official records. We thank present and former staff members of these establishments who have helped us, especially Dr W.L. Francis, a former SRC Secretary, J.P. Lonergan, formerly of the Australian Department of Science; and George Kazs and Kevin Bryant, Secretaries of the AAT Board during the period 1974–1979.

In Britain, Sir Bernard Lovell allowed us to draw on his research of the early history of the AAT which he published in the *Quarterly Journal of the Royal Astronomical Society*. Janet Dudley, former librarian at the Royal Greenwich Observatory, accommodated us for many hours as we researched the archives at Herstmonceux Castle. Colin Blackwell and Derek Fern, of Freeman Fox and Partners, provided much useful advice for the section on engineering. Also, we record with thanks our discussions with John Pope, Roderick Willstrop, Bernard Pagel, Paul Murdin and the late David Brown.

Gascoigne wishes to thank Don Mathewson and Alex Rodgers, successive Directors of Mount Stromlo and Siding Spring Observatories, for a Visiting Fellowship which enabled him to use the facilities at Mount Stromlo.

All the photographic plates used in the book are AAT Board material unless we have acknowledged them from other sources.

Finally, we cannot forget our families who patiently put up with us while we were producing this book in Sydney, in Canberra and in Farnham, Surrey.

S.C.B. Gascoigne, Canberra

Katrina Proust, Sydney

M.O. Robins, Surrey

1 The scientific background

1.1 Introduction

The end of the Second World War found astronomy in complete disarray. All over the world observatory staffs were scattered, their research programmes in abeyance, their telescopes unused for years. Certainly there was no suggestion that the subject was about to enter a new golden age, or that Britain and Australia would soon be the unquestioned leaders of the new science of radio astronomy; still less did anyone foresee that the same two countries were to build between them a major optical telescope which would be the envy of astronomers everywhere. Nevertheless the seeds for this regeneration were already contained within the war itself. Physics had played a notable and at times decisive part in the war, stood high in public esteem and was confident of continued government support. Again, the war had created a new breed of scientists who were used to responsibility, moved easily in the upper echelons of government and had forged strong personal links there. Perhaps even more directly, wartime radar led straight to radio astronomy, quasars and pulsars, wartime rocketry to satellites, X-ray sources and black holes. The brilliant discoveries which resulted, interwoven with and stimulating equally brilliant instrumental advances, established astronomy firmly in the front rank of the physical sciences and in the minds of the public. It was against this background that the Anglo-Australian Telescope was conceived, argued for, fought over, planned and built, finally to emerge as an astronomical triumph of the first order.

It takes many talents to build a large telescope; they run from high-level inter-governmental negotiations at one end to a maze of techniques from physics and technology at the other. In particular, a telescope makes extreme demands on many different kinds of engineering. Encompassing all this as it did, the building of the Anglo-Australian Telescope was a notable event with important scientific and political consequences. By far the largest optical telescope erected in either Britain or Australia, it has transformed optical astronomy in both. In Britain its success was a direct stimulus to the development of the northern hemisphere observatory on La Palma, the Observatorio del Roque de los Muchachos, and towards the building of the William Herschel Telescope, while in Australia it was a major force behind the building of the Australian National University's 2.3-metre telescope. It has had a substantial influence on large telescope design itself: it has set new standards for pointing accuracy, mechanical precision, and ease of operation, and the power of its control system has demonstrated clearly the practicability of the alt-azimuth mounting, which has been adopted for the designs of all recent large telescopes.

The first substantial moves towards building what became the Anglo-Australian

Telescope were made in about 1960, on both political and scientific fronts. An early step was to find funds for design studies. Then joint meetings were held, in 1963 in England and 1964 in Australia, at which many details were argued out and a formal proposal prepared which went eventually to the two Governments. Authority to proceed was finally received in April 1967. On the scientific side progress continued smoothly. Construction was entrusted to a Project Office, the organisation and staffing of which drew heavily on the experience gained with the large radio telescopes. Because of this the Office had particular strengths in structural and servo-engineering, expertise which had a considerable influence on the final design. Generous help was also given by the Kitt Peak organisation in the USA. Kitt Peak had been designing its own telescope for some time, and the original AAT design followed closely the original Kitt Peak design (but both were later modified substantially).

Neither the scientific case for the telescope nor its financial aspects gave rise to any real problems. On the political side, however, the story was different. One reason lay in differences between the systems for scientific funding and administration in the two countries. At that time, proposals in Britain were filtered through a complicated system of overlapping committees and advisory bodies, whereas in Australia there was, in effect, no system, and proposals like that for the AAT were considered *ad hoc*. Certainly matters in Britain were simplified by a major re-organisation of scientific administration in 1965, but the result was that the AAT then had to be considered as one of a number of large projects competing for money within the budget of the new Science Research Council.[1]

Also there was the inability of the astronomers to agree among themselves. For instance there were counter-proposals that the British join the European Southern Observatory, a consortium of European countries then in the process of formation, or that they upgrade the British Radcliffe Observatory in South Africa, or that they set up a telescope of their own somewhere in the Mediterranean area. At various times both sides considered joining forces with the Americans, while the Australians had to cope with strenuous opposition from their own biologists, and long silences from an apparently disinterested Government. Not surprisingly, the whole project was nearly abandoned on more than one occasion. Nor did the agreement of 1967 put an end to the difficulties, because shortly after that a long and bitter dispute broke out about whether the telescope would be operated through the Australian National University, or by an independent body answerable directly to the Anglo-Australian Telescope Board. Eventually all was settled, in a sense by the telescope itself. Once it went into regular operation, it worked so well that the differences disappeared in a wave of euphoria. Now they are but a distant memory.

1.2 The state of British astronomy
At the end of the Second World War, British theoretical astronomy stood as high

as ever. On the other hand, British observational astronomy had sunk to a nadir, so much so that at a function held in 1971 in honour of Sir Richard Woolley as retiring Astronomer Royal, one of the principal speakers, Sir John Carroll, was constrained to refer to *'the superannuated scrap heaps that constituted most of British Astronomical equipment when we were both young men, or even to a great extent at the time of his* [Woolley's] *appointment as Astronomer Royal'*.[2] It seemed a long time since the nineteenth century days when Huggins and Lockyer played central roles in the development of astronomical spectroscopy, or when De la Rue, Roberts and Common were among the leading pioneers of astronomical photography. This too was the period when most of the world's large reflecting telescopes originated in Britain. Indeed, when John Herschel said *'that eminently British instrument, the reflecting telescope'*[3] William Herschel, the Earl of Rosse, Lassell, Grubb and Common were building in succession telescopes in the 36- to 72-inch range. Grubb's largest telescope, a 48-inch, went to the city of Melbourne in 1869; we refer to it later.

Reasons can be adduced for the decline in British optical astronomy – the inadequacies of the British climate, new directions in astronomy, advances in telescope technology. In the same way observatories lost ground in both northern Europe and eastern USA. The trend continued right up to the Second World War, and we pursue it further in the next chapter. At the end of that war, Redman described the difficulties at Cambridge in these words: *'We found ourselves with no modern equipment whatsoever and with a great shortage of staff, in a country extremely disorganised by war, at a time when new telescopes or other apparatus were a matter only of rather distant hope, and when there were very few young astronomers, for hardly any had been trained for six or seven years.'*[4] It was widely recognised that physics had made an enormous contribution to the war, and in the post-war era support of university physics departments by the Department of Scientific and Industrial Research (DSIR) increased dramatically, to five and ten times the pre-war figure.[5] But with the notable exception of the Isaac Newton Telescope, astronomical support went mostly to radio and space astronomy. It must have been difficult to find anything for those old telescopes to do that could not be done better elsewhere; perhaps also the subject was perceived as still living in the shadow of the enormously successful nuclear physics and quantum mechanics of the previous decades. The most telling commentary was that made by the young astronomers who voted with their feet, and joined a mass exodus to the American observatories. Britain lost much of its most promising astrophysical talent in this way, including E.M. and G.R. Burbidge, A. Dalgarno, J. Faulkner, R.H. Garstang, T. Gold, T.D. Kinman, W.L.W. Sargent, L.E. Searle, Malcolm Smith, P.A. Strittmatter and N.J. Woolf. Five members of this group were subsequently elected Fellows of the Royal Society of London.[6]

Some encouragement had been given in 1946, when impressed by the parlous state of British observational astronomy, a group of senior scientists approached the Government for funds to build a large telescope in Britain. Within six months the money was forthcoming for a 100-inch, the Isaac Newton Telescope.[7] But 21 years were to pass before it was actually built. One cause of the delay was a long, drawn-out dispute over the optical design; another, as Graham-Smith and Dudley point out, was *'that the designers were severely handicapped by the very condition that the telescope itself eventually helped to cure, namely the almost total lack of an essential team comprising both experienced observers and telescope designers'.*[8] There had been nowhere for them to learn.

Another lengthy dispute went back to 1930. In that year the Radcliffe Trustees, who were responsible for the Radcliffe Observatory in Oxford, sold a parcel of land in England and proposed to apply the proceeds to establish a 74-inch telescope and observatory in South Africa; the municipality of Pretoria had offered a suitable site. A group headed by Professor F.A. Lindemann of Oxford (later Lord Cherwell) opposed the move, arguing that the assets of the trust should not be moved outside the jurisdiction of the English Courts, and therefore that the telescope should be located in England. The case went to the Court of Chancery, where judgment in favour of the original proposal was not given until 1934.[9] Because of this delay the construction of the telescope was further held up by the war, and only in 1948 did it finally go into operation. The success of the Radcliffe Telescope is well-known. With a total staff of four and a half (later increased), it has been described as the most cost-effective telescope ever built. An arrangement was made whereby a third of its time was allocated to observers from the Cape Observatory, effectively the southern arm of the Royal Greenwich Observatory. The arrangement was later extended by Woolley to include observers from Britain, with the aid of grants from the DSIR. The experience must have given many British astronomers their first real taste of observing.

Woolley took over as Astronomer Royal late in 1955, when the Greenwich telescopes were being recommissioned at their new home at Herstmonceux. He made a great contribution to British astrophysics. Besides the South African programme just described, he quickly settled the disputes which had been holding up the Isaac Newton Telescope; he organised regular observing sessions on the 74-inch at Helwan, Egypt; he brought new astrophysics programmes, new ideas and new people into Herstmonceux; and he established a large new workshop there, on the lines of that of his old observatory at Mount Stromlo. But it could well be that he made his most important contribution when, virtually single-handed, he launched the AAT project and gave it the initial impetus so essential to its success.[10]

In this context we should mention Professor R.O. Redman, whose small observatory at Cambridge became noted for innovative programmes which could be run effectively despite the English climate. Besides much work on the sun, these included some of the earliest work on infrared observations with lead sulphide detectors, on Fourier spectroscopy, on photoelectric narrow-band spectroscopy, and included the invention of a quick, accurate method for measuring radial velocities (the Griffin device) which has been widely adopted elsewhere.

This section has been confined to optical astronomy, and, if it were the weakest astronomical link, that very weakness was an argument for building the AAT. But the phenomenon which made the indelible impression on the post-war period was undoubtedly radio astronomy, to which we return later.

1.3 The state of Australian astronomy

In a memorable passage, a noted astronomical historian has described James Cook's discovery of the eastern seaboard of Australia as 'technological fallout from an international astronomical endeavour'.[11] Be that as it may, Cook's first voyage, and his observation of the 1769 transit of Venus, are firmly implanted in the Australian consciousness, and have given astronomy in that country a cachet which survives to this day. Astronomy is not only the oldest Australian science, but also one of the strongest.

The first 'permanent' observatory in the southern hemisphere was established in Australia, by Governor Brisbane at Parramatta in 1821. A soldier of distinction, Brisbane was also an enthusiastic amateur astronomer. It was said that one of his principal reasons for accepting the Governorship was the opportunity it gave him to observe the little-known southern sky. The British Government, already committed to an observatory at Cape Town, would not support his venture, and he had to finance instruments, books and staff from his own pocket. He saw his main task as mapping the southern stars, and the *Parramatta Catalogue* of 7385 stars, published in 1835, was well-known in its time. Unfortunately the telescope with which the observations in right ascension were made was faulty, and with the passage of time the catalogue has become discredited. Following Brisbane's return to Europe in 1825, the observatory declined and gradually fell into disuse.

However, it was not long before a need arose for services like those provided by the Royal Greenwich Observatory in England. In particular, ports as remote as those in Australia found it essential to have accurate time for checking ships' chronometers. Observatories were accordingly set up in Victoria in 1853, New South Wales in 1856, South Australia in 1874, and Western Australia in 1896. They were responsible not only for providing time, but also for primary trigonometric surveys, meteorological services, tidal and magnetic data, and weights and measures. Like Greenwich, they were the scientific workhorses of their respective Governments, and their directors were men of standing.

What would now be called research, and cultural activities generally, were by no means discouraged. The most striking example of this policy was provided by the Great Melbourne Telescope. It was a 48-inch reflector acquired by the city of

Melbourne in 1869, in the wake of the Victorian gold rush. From 1870, when it went into operation, until the end of the century it was the biggest equatorially mounted telescope in the world. But for reasons explained in the next chapter, it was a failure. In an oft-quoted remark, G.W. Ritchey claimed it held back the development of large telescopes for 30 years. When the Observatory closed, it was acquired by Mount Stromlo, where modified and refurbished, it at last came into its own and gave good service for many years.

At the turn of the century there began a long decline which saw the demise of all the State Observatories except Perth. They were hard hit by the depression of the 1890s, especially in Victoria, and fared little better in later depressions. They lost many of their civil functions; surveying, for instance, went to Lands Departments, and the observatory time services were made redundant by radio time signals. Worst of all the Commonwealth Meteorological Bureau, established under the new Australian Constitution, took over meteorology from the State Observatories in 1908, and at one stroke deprived them of their most visible means of support. They lost half their staff, half their income and much of their public standing: it was a blow from which they never recovered.

What was left with the State Observatories was the Australian share of the Carte du Ciel or Astrographic Catalogue, an ambitious programme adopted internationally in 1887 (all too hastily, as one historian put it)[12] for the photographic mapping of the whole sky. With a population of barely two million, Australia was made responsible for three of the eighteen zones, a greater area than given to any other country except France. This immense task dominated official astronomy in Australia for decades, and effectively excluded the prosecution of any other work. Final measuring and cataloguing was not completed until 1964 at Sydney Observatory, by which time the Adelaide and Melbourne Observatories had already been closed. Sydney, which survived so long only because of the quality of one of its later directors, Harley Wood, was itself closed in 1982.

The remarkable post-war resurgence of Australian astronomy owed little to the State Observatories: its origins lay rather in two Commonwealth bodies, the Commonwealth Solar Observatory (later the Mount Stromlo Observatory),[13] and the Radio Research Board. The Solar Observatory was founded in 1924. Like the Meteorological Bureau it was a Commonwealth and not a State body. Its founder and first director, Walter Duffield, died in 1929, and a ten-year interregnum followed which ended when Richard Woolley was appointed his successor in 1939. Woolley was a product of the classical British school of Cambridge and Greenwich, and his appointment was critical, because he came with the express purpose of discontinuing the existing programmes of solar and geophysical work, and of redirecting the Observatory solely into stellar astronomy.

Woolley's plans were disrupted by the outbreak of the Second World War. Instead, his Observatory became a munitions factory, with a peak staff of about seventy engaged mostly in the design and manufacture of precision optical instruments for the armed forces. It proved to be an admirable, and among optical observatories, a unique training for what lay ahead. At the end of the war Mount Stromlo found itself in much the same straights as the British observatories. Of its four telescopes only one, a 30-inch reflector, had been built this century. All were run-down, little used, and badly in need of overhaul. The 48-inch Melbourne Telescope, acquired as scrap in 1945, was in considerably worse condition, and even a new 74-inch, received from its British makers Sir Howard Grubb, Parsons and Company in 1955, had its problems in the shape of an astigmatic primary mirror. However, the Mount Stromlo workshops were bigger, better staffed and altogether more competent than before the war, and with the confidence gained from years of making precision military instruments proved fully equal to the tasks ahead. Once the telescopes were in working order, good programmes for them were easy to find. With their technical backing, good climate and access to the southern skies, they could be used to much better advantage than would ever have been possible in Britain. By the time Woolley returned to England in 1955, some noteworthy papers had been published and the young institution was beginning to make a mark.[14]

The 74-inch was important. It is a sister instrument to that at Pretoria, and for 20 years the two telescopes were the largest in the south. It was through experience gained with the 74-inch that Australian optical astronomy came of age, in the sense that it was brought face to face with urgent contemporary problems. As has been said elsewhere

> . . . much that was learnt with [the 74-inch] later went into the Anglo-Australian Telescope. In fact without the 74-inch there might not have been an AAT. That such a good case could be made for the AAT came directly from the experience in installing, maintaining, and above all in observing with the 74-inch.[15]

. . . and, it should be added, in being brought up against its limitations.

The Radio Research Board (RRB) was established in 1926, initially at the instance of the Universities of Sydney and Melbourne; it operated within the framework of the Council for Scientific and Industrial research (CSIR). It was the spearhead of the strongest branch of physics in the country, the only one which up to that time had gained convincing international status. When the war came and steps were taken to set up a radar laboratory in Australia, the RRB was the natural starting-point. From the outset it was agreed that, because of the outstanding work of the CSIR–RRB physicists, the new laboratory would not merely duplicate existing British equipment but would carry out its own research and development. This was a most important decision. It led to the flowering of some fine talents, and paved the way for the equally important decision that after the war the laboratory would remain in being as a permanent division of the CSIR, working on peace-time applications of radio science. The Radiophysics Division contributed notably to cloud physics and to aircraft navigation and landing systems, but it made its real name in radio astronomy.

1.4 The radio astronomers

The origins and early years of radio astronomy have been described by many people and already constitute something of a classic in the history of science. We content ourselves here with those parts of it which concern us directly. Radio astronomy developed on closely parallel lines in both Britain and Australia. This is no surprise when one recalls that the people involved – ex-radar physicists, engineers and servicemen – came from very similar backgrounds, with similar training and experience, and used similar equipment. They also began from similar starting-points. These stemmed not so much from the pioneer work of Karl Jansky and Grote Reber in the USA as from discoveries made by J.S. Hey, initially when he was working with the Army Operational Research Group during the war.

Hey had found first that the sun was a powerful and highly variable source of radio noise; then in 1946 he discovered a localised source in the constellation Cygnus, possibly a quite compact one. Observations of the sun were taken up shortly after the war by Ryle's group in England and Pawsey's in Australia. Both sides worked with adapted wartime radar equipment, and both used interferometers, the first moves in a long battle to secure adequate resolution at radio wavelengths. The solar work led naturally to observations of the Cygnus source, and before long the discovery of further discrete sources had ushered in the new science of radio astronomy. The study of the sun was not taken far in England, but in Australia it became a considerable part of the radio astronomy effort. Within a few years J.P. Wild and his team were undisputed leaders in the field, and had transformed our ideas of the solar corona.

Initially the people from the radar ranks knew nothing of astronomy. They were able young physicists and engineers who were not only highly skilled in the most advanced electronic techniques of their day, but had acquired habits of work and, in Lovell's words, an outlook on research

> . . . utterly different from that deriving from the pre-war environment. The involvement with massive operations had conditioned them to think and behave in ways which would have shocked the pre-war university administrators. All these facts were critical in the large-scale development of astronomy.[16]

The influx of so many new faces and new ideas, and the freshness of approach that accompanied them, made a deep impression on the optical astronomers, already shaken by the knowledge that, left to their own devices, it would have been years before they hit upon the real significance of radio galaxies. However, within a decade the two disciplines had begun to merge, so much so that the needs of the radio astronomers for optical identifications of their sources became one of the most compelling arguments for the new telescope in the south. Only with such an identification could an optical spectrum be obtained, and optical spectra were all-important as the only source of radial velocities and hence of distances. In addition, they provided unique information about physical conditions within the sources themselves.

Radio astronomy also had much to contribute to the engineering side of astronomy. As pointed out earlier, engineers who had cut their teeth on huge steerable radio telescopes like the 210-foot at Parkes, New South Wales, or the 250-foot at Jodrell Bank, Cheshire, had something new and valuable to bring to big optical telescopes like the AAT.

For physics at large, the most influential of the earlier discoveries of the radio astronomers were those of extragalactic radio sources and of quasars or quasi-stellar objects (QSOs). We describe here three major contributions; all sent waves almost of disbelief running through the astronomical community, and all figured prominently in proposals for large telescopes. In 1949 Bolton and Stanley suggested optical identifications for three of several 'radio stars' they had discovered with the cliff interferometer at Dover Heights, Sydney. The source, Centaurus A, was identified with the object NGC 5128, and the source, Virgo A, with NGC 4486; both NGC objects were soon shown to be external galaxies with unique features. The third identification, the Crab Nebula, was local, a supernova remnant within our own galaxy. This was a tremendous event, the advent of a whole new astronomical species – two species perhaps – with properties at once completely unexpected and of the greatest significance.

A little later, Graham-Smith in Britain determined an accurate position for Hey's Cygnus A, which Baade and Minkowski using the Palomar 200-inch were then able to identify with a pair of faint, apparently interacting galaxies, indubitably extragalactic. This was in 1954. The surprise was the tremendous excess of the radio power of their object over its optical power; here was the second brightest radio source in the sky, yet the world's largest telescope was required to find its optical counterpart.

The third contribution is probably the best known. Observations made with the long base-line interferometers at Jodrell Bank had drawn the attention of another British astronomer, Cyril Hazard, to the extremely small angular size of the source 3C 273. Using the 210-foot radio telescope at Parkes, he was able to determine an accurate position by observing it as it was being occulted by the moon. With this position it was identified unambiguously with an apparently nondescript thirteenth magnitude star. But a spectrum, again taken with the Palomar 200-inch, showed the 'star' to be a remote galaxy receding at a sixth of the velocity of light, and therefore at a great distance. This was the first quasar. Quasars are now numbered in thousands. They are the most luminous and the most distant objects known, and so provide the best available tools for studying the morphology and the outer reaches of the Universe. Further, the problem of how their enormous power output is generated remains one of the most challenging in contemporary physics. The existence of quasars alone would have justified building a large southern telescope.

If the influx of radio astronomers changed the face of optical astronomy, in the 1960s a second comparable influx accompanied the development of space astronomy, as conducted first from rockets, then from satellites. As in radio astronomy, the initial impetus came from a need to learn more about the sun's radiation, if only to allow the physics of the earth's ionosphere to be more fully understood. This of course was at ultraviolet and X-ray wavelengths. The successful observation of the sun, which was of great importance in itself, was followed by the discovery of other X-ray sources, found at first by chance, then by systematic surveys. Out of these there developed a major new branch of astrophysics. It was too late to have much effect on the early stages of the AAT, although within a few years it would have strengthened the scientific case considerably.

1.5 The scientific case

The astronomical community had long been aware of the pressing need for a large telescope in the south. While all the existing large telescopes were well north of the equator, an undue proportion of the really important astronomical objects, some of them crucial to the further development of the subject, were in the southern third of the sky, out of their reach. Those objects included the galactic centre, already known to be complex and enigmatic, and which at Australian latitudes passed directly overhead; the southern and brightest third of the Milky Way; the Magellanic Clouds, the two galaxies nearest our own, nearer than the Andromeda Nebula by a factor of ten; the most accessible globular clusters, the ancient survivors of the earliest phase of our own galaxy; the great, virtually untouched class of southern galaxies; innumerable other special and peculiar objects of all kinds; and of course, the southern radio sources.

It was hoped that through them the AAT would contribute to solving the major astronomical problems of that time: the origin and evolution of stars and galaxies, the synthesis of heavy elements in stellar interiors, the dynamical structure of the Milky Way, cosmology and the extragalactic distance scale, the physical nature of the radio sources. To emphasise the point a provisional, explicit programme on these lines was drawn up, on which, had the telescope been available, work could have begun at once and continued for years.

A word should be said about the special importance of the Magellanic Clouds.

Besides being so near, the Clouds are exceedingly rich in a great variety of objects – aggregates of stars of all types and ages, hydrogen emission regions, supernova remnants in all stages, and so on – all at effectively the same distance, and most of them free from the obscuring dust clouds which are such a complication in our own galaxy. The Clouds present an ideal ground for calibrating distance indicators. The demonstration by the Pretoria observers that, period for period, the Cloud RR Lyrae variables are only a quarter as bright as the Cloud delta Cephei variables was an early example, most important for the extragalactic distance scale. The Clouds are galaxies in their own right, of a type common in our part of the Universe, and they should contain data enough for us to be able to reconstruct their evolutionary histories in detail. Being so near to each other and to our own galaxy, they constitute a system of interacting galaxies, again of a common type, which we are allowed to study at the closest of quarters. They are truly the Rosetta Stones of astronomy.[17]

That was in 1964. Those problems remain, but within ten years, with the telescope not yet complete, the picture had been transformed from that of a placid, rather stately Universe of deliberately evolving stars and galaxies to one riven, in Owen Gingerich's words, by *'the cataclysmic explosions of supernovae and galactic nuclei, the phenomenal outbursts of quasars and radio binaries'*:[18] a violent Universe indeed. The old, 'thermal' Universe of 'normal' stars and galaxies was driven by the thermonuclear conversion of hydrogen into helium, and Newtonian gravitation. The new Universe is the domain of high-energy astrophysics, of radio galaxies, quasars, supernovae, pulsars, jets, X-ray sources and the Big Bang. In it, general relativity reigns supreme. It in turn is driven by the synchrotron radiation emitted by highly relativistic electrons, the gravitational energy released by matter accreting onto the superdense neutron stars and black holes, and the awesome supernova explosions, all highly relativistic processes. These are the phenomena, unimagined before the days of radio and space astronomy, which have created such an impression on modern physics.

There has been a second revolution, less spectacular than that above, but in many ways of comparable importance. The photographic plate, the staple detector in astronomy since the turn of the century, has been supplanted by various types of digital electronic devices which have raised the efficiency by which light is measured 20-fold or more. Moreover, the new devices have additional advantages in that their response is linearly proportional to the incident light, and that they can feed their data directly into computers. The result is that a 40-inch telescope with the instrumentation of the 1980s could in most respects out-perform a 150-inch instrumented in 1964, and observations then seen as being at the very limit of the AAT are now commonplace. So not only can the AAT work on a range of problems far wider than anything envisaged in the original submission, but it can do so many times more efficiently.

Notes to Chapter 1

1 Lovell A.C.B. (1985). *Qrt. J. R. Astr. Soc.* **26**, 415

2 Carroll Sir John (1972). *Qrt. J. R. Astr. Soc.* **13**, 132

3 Quoted by Clerke A. (1893) in *A Popular History of Astronomy during the Nineteenth Century*, 3rd ed., p.150

4 Redman R.O. (1960). *Qrt. J. R. Astr. Soc.* **1** 10 'Presidential Address'

5 Lovell A.C.B (1987). *Qrt. J. R. Astr. Soc.* **28**, 1

6 The five were E.M. Burbidge, G.R. Burbidge, A. Dalgarno, T. Gold, W.L.W. Sargent.

7 Woolley R.v.d.R. (1962). *Qrt. J. R. Astr. Soc.* **3**, 249 '. . . the Council of the [R A] Society took the matter up with vigour. Events thereafter moved with astonishing speed. Within six months the British Government had accepted the Society's main submission . . .'

8 Graham-Smith F. and Dudley J. (1982). *J. Hist. Astron.* **13**, 1

9 (1934) *Observatory* **57**, 250

10 McCrea W.H. (1988). 'Richard van der Riet Woolley' *Biog. Mem. Fel. Roy. Soc.* **34** and *Historical Records of Australian Academy of Science* **7**, 315

11 Gingerich O. (1976). *Vistas in Astronomy* **20**, 2

12 Pannekoek A. (1961). *A History of Astronomy*, Allen & Unwin, 337

13 Later again it was renamed the Mount Stromlo and Siding Spring Observatories (MSSSO). In this book we refer to it mostly as Mount Stromlo or MSO.

14 Gascoigne S.C.B. (1984). *Proc. Astron. Soc. Aust.* **5**, 597 'The Woolley Era'

15 Gascoigne S.C.B. (1988). *Australian Science in the Making* R.W. Home (ed.) Cambridge University Press, 358

16 Lovell A.C.B (1987). *Qrt. J. R. Astr. Soc.* **28**, 8

17 Aller L.H. (1963). *Science* **139**, 21

18 Gingerich O. (1976). *Vistas in Astronomy* **20**, 8

2   The technical background

2.1 The early history

The first refracting telescope about which we know anything definite was the one Galileo directed at the sky in 1609; the observations he made with it established the Copernican heliocentric theory and changed the course of western civilisation. The first reflector was made by Isaac Newton in 1668. It had a speculum mirror with a diameter of one and a third inches and a focal length of six inches. Dolland made the first achromatic objective in 1767. Throughout this period telescopes improved slowly, always hampered by the limitations of technology, difficulties with materials – suitably homogenous glass for refractor objectives, suitable metal alloys for the mirrors of reflectors – and by the scarcity of skilled artisans. It was not until the nineteenth century that advances in glass technology enabled reasonably large (8- to 10-inch) refractors to be built, and not until after mid-century that the discovery of a process for depositing silver films on polished glass made reflectors practicable for general use.

During the early 1800s reflectors reached something of a pinnacle, largely in the hands of William Herschel, who with their aid made a series of brilliant discoveries which laid the foundation of stellar astronomy. But for the rest of the nineteenth century it was refractors which were the preferred instruments of professional astronomers. They were the instruments of precision, lending themselves to accurate measures of star positions and to observations of double stars and stellar parallaxes in a manner reflectors could not emulate. They became the source of almost all accurate data, especially of the enormous catalogues of star positions which were such a feature of the time. Towards the end of the century, what was virtually an international competition developed to build the largest refractor. In quick succession, larger lens blanks and objectives were produced, culminating in the 36-inch erected at the Lick Observatory near San Francisco in 1888, and the 40-inch erected at Yerkes Observatory near Chicago in 1897. These, together with comparable instruments built for Meudon Observatory near Paris, and Potsdam Observatory near Berlin, were well-engineered precision telescopes which have remained in use until the present day.

The Yerkes refractor was, however, the end of the line. Formidable obstacles stood in the way of making larger lens blanks of the requisite optical quality, and of mounting the finished lenses when made. Further, refractors could not be made fully achromatic, that is, they could not bring the full range of colours to a common focus, and they could not be made fast enough for the effective photography of nebulae. On the other hand, mirrors could be made of inferior glass which did not even have to be transparent, and because of this blanks for mirrors could be made much larger. But the real reasons why refractors were ultimately superseded by reflectors went deeper. The first was that in 1860 Kirchhoff and Bunsen arrived at a new understanding of the physics of radiation which led directly to spectroscopic analysis and astrophysics. The second was the steady advance of astronomical photography, especially after the introduction of fast dry plates in 1875. The impact of photography was tremendous, and it was no exaggeration to claim that 'photography . . . effected a revolution in Astronomy equal to the invention of the telescope'.[1] Not only could a photographic plate reach much fainter than the human eye, but it could record a multitude of images in one exposure, with those exposures preserved in permanent form for subsequent study. Finally, while spectroscopy had opened endless new possibilities for the study of astronomical objects, it was photography which provided the practical means which enabled those possibilities to be realised. Between them, spectroscopy and photography transformed the whole subject.

Equally, they transformed the telescope which had to undergo sweeping changes before it could record this new kind of information. Initially, it was not clear whether the new demands could best be met by refractors or reflectors. At that time reflectors were the province of wealthy amateurs. Their advantages were well recognised: they were achromatic, they were cheaper, they could be made with low focal ratios, and so with high photographic speed, and there seemed no obvious limit to the size to which they could be built. On the other hand, for much of the century they were awkward unmanageable instruments, often crudely engineered, unusable for photography, and capable only of a limited range of visual observations. Major improvements, which it took several decades to implement, had to be made before reflectors could achieve general acceptance.

In this process a key role was played by an English telescope. In 1885 a 36-inch reflector, built by Common in London in 1879, was sold to Edward Crossley of Halifax, but despairing after some years of the climate of his native Yorkshire, Crossley presented it to the Lick Observatory in California, where it was erected in 1895. There it was taken over by the incoming director, James Keeler, who had to spend two years modifying it before he could at last obtain with it a series of pictures of nebulae which quite surpassed anything made previously, and which for the first time revealed to astronomers the true richness of the extragalactic Universe.

This success, coupled with that of a 24-inch reflector built by Ritchey at Yerkes Observatory specifically for photography, led directly to the construction of the Mount Wilson 60-inch. The 60-inch was a tremendous advance, eclipsing the giant Lick and Yerkes refractors in almost every detail. Thereafter, given the stimulus of some excellent mountain sites, a climate immeasurably better than that in Britain or anywhere else in north-west Europe, and the generosity of a series of wealthy individual patrons, the centre of observational astronomy shifted decisively to California. From

1900 to 1950 seven reflectors between 60 and 200 inches aperture were built in the USA, one of them, a 72-inch, for the Dominion Astrophysical Observatory in Canada. Over the same period the largest telescopes built for British observatories were two 36-inch reflectors, one in 1929 for Edinburgh, the other in 1935 for Greenwich.[2] It should be stated, however, that during that period Sir Howard Grubb, Parsons and Company of Newcastle–upon–Tyne built two 74-inch reflectors, one for Toronto in 1935, and a similar one for Pretoria in 1948. The third of the series, the 74-inch for Mount Stromlo, was completed early in the 1950s.

The success of the Mount Wilson 60-inch inaugurated a new era in the history of the telescope. As Horace Babcock said of it, *'this telescope, at an excellent site, was constructed to much higher standards of optical and mechanical engineering than had been attempted before. The instrument was unusually massive and rigid, yet with a mechanism that was smooth and precise'.*[3] It led directly to the 100-inch, but although the feasibility of the latter was already under study before the 60-inch had been completed, in 1908, the 100-inch was not in regular operation until 1919. In turn the 100-inch inspired the 200-inch, a magnificent instrument which went into operation in 1949, to dominate the large telescope world for 20 years. The AAT may be regarded as one of its direct descendants.[4]

2.2 The Palomar 200-inch Telescope

The main requirements of a large telescope are well-known. The mirrors, of which the largest in the AAT has a diameter of 150 inches, must be figured to the very small tolerances demanded by the theory of light, and supported so that they retain their true shape no matter in what direction the telescope is pointing. The mounting must carry the mirrors, with their associated instruments – spectrographs and the like – in accurate collimation, again regardless of where the telescope is pointing. Third, the telescope must be able to set and track on a star to a fraction of an arc second for appreciable lengths of time. The whole should be reliable, and quick and convenient to operate.

The important consideration is how rapidly the difficulty of meeting these requirements increases with the size of the telescope. We illustrate with two particular cases, the first the primary mirror. A horizontal disc-like mirror supported around its circumference will sag under its own weight by an amount proportional to $(\text{diameter})^4/(\text{thickness})^2$. To produce a perfect image the mirror should maintain its shape to within a tenth of a wave, regardless of its size. Given a mirror just on this limit, suppose we want to scale it up by a factor of two. The thickness and diameter will both be doubled but the mirror itself will sag by four times as much, and to keep it within the tenth-wave limit either the thickness must be doubled again, and hence the weight increased 16-fold, or the support system which enables it to maintain its shape must be made to work four times as well.

The second example is the pointing and drive accuracy. For a precision laboratory instrument or a surveyor's theodolite, the sub-arc second accuracy referred to above is not impossible to meet. But with a telescope like the AAT the moving parts weigh about 265 tonnes; they will distort appreciably under their own weight and will exert considerable pressure on their supports. Such problems call for an extremely stiff and stable structure, with drive components – motors, gears, bearings – of the highest quality, and movements with the lowest possible friction. The mass of such a structure and the general difficulties of its design increase rapidly with the size of the mirror. It is these great masses, building them and moving them with the precision expected of fine instruments, that create the special problems of large telescope design, and bring their construction to the limits of engineering practicability.

These problems were well-known, but the Palomar 200-inch (now known as the Hale 5-metre Telescope) was so great an advance over the Mount Wilson 100-inch that in its particular case the problems became almost new in kind. The giant Palomar Telescope, one of the supreme engineering feats of the century, was completed in 1949 and dominated the large telescope world for the next 20 years. Though conceived as a logical development from the 100-inch, it was many times heavier and more ambitious, and its sheer size created problems of a new dimension. That they were solved so resourcefully is part of astronomical history, and the solutions found have long since been standard practice in telescope technology.

Thus the 200-inch was the first telescope to use a horseshoe yoke to carry the tube and mirrors, and it was the first to support the polar axis on virtually frictionless oil-pad bearings. The choice of what was then the ultra-fast focal ratio of f/3.3 for the primary mirror broke new ground, and the Serrurier truss, devised to answer the critical alignment problems created by these new optics, must have been used for the tube of almost every subsequent telescope. It was the first telescope in which the observer could ride at the prime focus inside the tube, and, although not generally acknowledged, it was the first to use a so-called 'thin' mirror.[5] These innovations were the hallmarks of the 200-inch, and one way and another they have been used in every sizeable telescope built since, the Kitt Peak and Anglo-Australian telescopes not excepted. Except for the Russian alt-azimuth, all the telescopes of the AAT period are variations on the 200-inch theme, modified according to various perceptions of how best to take advantage of advances in engineering methods, materials and technology generally.

2.3 A new wave of telescopes

We described in the preceding chapter how the exciting astronomical discoveries of the post-war years and the great surge of enthusiasm they engendered in the scientific world led to the proposal to build the AAT. This excitement was not of course confined to Britain and Australia. When the AAT was approved in 1967, similar proposals had already been made in Canada, Europe and the USA, in the latter two for large telescopes in the southern hemisphere. In the next few years proposal followed proposal, and the resulting state of affairs, which contemplated something like a five-fold increase in the number of major telescopes, at an aggregate cost of hundreds of millions of dollars, was without precedent in the long history of the subject.

The proposals which have come to fruition are listed in Appendix 9. Note that apart from the Russian 6-metre and the Keck Telescope, both special cases in several ways, all the telescopes listed are smaller than the 200-inch. In fact their sizes are curiously similar, most of them falling within the 3.5- to 4-metre range. Perhaps it was that the 200-inch had become the benchmark, and that with up-to-date engineering one could expect a 150-inch to at least equal the performance of that now elderly telescope, at about half the cost. And if the new electronic detectors were to realise only a fraction of their promise, the newer telescopes would perform considerably better. Another reason was that it would certainly take longer, perhaps much longer, to obtain a mirror blank for a 200-inch than for a 150-inch. And while the Palomar (Hale) Telescope remained the obvious model for its successors, its technology went back to the 1930s, and no modern engineer would do more than follow it in broad outline.[6] As a recent writer has put it:

> The Hale [telescope] took twenty-one years to build – from 1928 to 1949. It contains thousands of components – motors and relays, gears and wheels, pipes and pumps – dating from the nineteen-thirties: parts made by companies now bankrupt or merged, parts unobtainable, parts no longer understood. Colossal, aloof, agile, seemingly indestructible, and magnificently extragalactic, the Hale telescope stands among all telescopes as the climax of dreadnought design.[7]

Three of the telescopes in Appendix 9 went to the southern hemisphere. In order of commissioning they were the AAT; the Cerro Tololo Inter-American Observatory's (CTIO) telescope in Chile, the southern counterpart of that on Kitt Peak; and the ESO telescope. The European Southern Observatory (ESO) is a consortium of European nations[8] set up to establish a multi-national observatory in the southern hemisphere. It too selected a site in Chile, on the southern edge of the Atacama Desert and near the CTIO. The concentration of optical telescopes in this part of Chile is rivalled only by that on Mauna Kea.

The New Astronomy was greeted with quite as much enthusiasm in the USA as elsewhere, and before long the number of American astronomers had grown to such an extent that the demand for telescope time far exceeded the supply. The problem was exacerbated by the fact that all the major telescopes were in 'private' observatories where the resident staffs had absolute first priority on their use. The rapidly growing importance of astronomy and critical nature of the observing position demanded that some action be taken. In May 1960 a new federally funded organisation therefore was set up with its headquarters in Tucson, Arizona. This was the Association of Universities for Research in Astronomy (AURA). Its function was to build and to operate telescopes, primarily for American users, and its operational arms were to be the Kitt Peak National Observatory (KPNO) in Arizona, the Sacramento Peak Solar Observatory in New Mexico, and later the CTIO in Chile.

KPNO was to include a number of telescopes, the largest being a 150-inch. This figure was a compromise. The telescope had to be big enough to attack front-line problems, and to carry an observer in a prime focus cage, 200-inch fashion – in fact, to be thoroughly competitive with the 200-inch – but account also had to be taken of the time and cost factors.[9]

In May 1963 KPNO commissioned the Westinghouse Electric and Manufacturing Company to carry out a design study for its mounting. Thirty years before, Westinghouse had been a prime contractor for the 200-inch, and it was natural to start where, in effect, the 200-inch had left off. This study was the direct basis for the design of the Kitt Peak telescopes; AURA ordered a second, identical telescope for CTIO in Chile. KPNO set up a Project Office in Tucson, and by the time the AAT was approved in April 1967, it had made considerable progress on the design concept of its own telescope. This became also the basis of the design for the AAT.

2.4 Early design studies

The first step towards a technical specification for the telescope was taken when Woolley convened a meeting, primarily of astronomers, at Herstmonceux in October 1963. Earlier in the year he had approached the British Department of Scientific and Industrial Research for a grant for the purpose. One decision was for a symmetrical equatorial mounting, as opposed to an alt-azimuth. Support for an alt-azimuth had been gathering strength, especially in Britain, but several years had to pass and some progress made with the AAT before its proponents could produce a sufficiently convincing case.[10] Another decision was that the mirror diameter should be 150 inches, not the 120 inches proposed originally. The arguments were similar to those used for the KPNO telescope, with perhaps more weight given to the time factor: the British were acutely aware of the seventeen years that had already elapsed since the beginning of the Isaac Newton Telescope (INT), and Mount Stromlo's 74-inch had taken seven years to complete, even though it was built to an existing and well-tried design.

A second meeting was held in Australia in June and July 1964. It was attended by R.v.d.R. Woolley, A. Hunter, J.D. Pope and R.O. Redman from Britain, and by J.C. Bolton, S.C.B. Gascoigne, A.W. Rodgers and H.P. Wehner from Australia (Bok was overseas), most of whom played leading roles in subsequent AAT operations. They inspected prospective sites – Mount Singleton in Western Australia, Mount Serle in South Australia, and Siding Spring Mountain in New South Wales – and the radio telescope at Parkes, but their main task was to draw up a complete specification of the telescope, and indeed of all the technical aspects of the submission. After it had received the blessing of the Royal Society of London and the Australian Academy of Science, this became the formal submission which was presented to the respective Governments on 30 July 1965.[11]

The optical specifications were quite detailed and in the event were adhered to closely. One figure which reads curiously in the light of subsequent events was that for the setting accuracy: it was to be within '30 arc seconds of non-repeatable error'. Actually 10 arc seconds had already been suggested, and remained the target figure. Emphasis was given to computer control, with specific mention of corrections for refraction and for mounting and tube flexure. Not much was said about the mounting, though the possibility of using the KPNO design, work on which could hardly have begun, was already in the air. The choice of site, as between those mentioned above, was left open. Thereafter no further technical work was carried out until May 1967.

To the astronomical community the submission produced no discernible reaction from either Government, and hope slowly ebbed away. It was startling to learn, out of a blue sky, that it had after all been approved. *'Suddenly risen from the dead'*, as Redman wrote to Vincent Reddish.[12]

The interim agreement is discussed in Chapter 4. It contained a number of clauses, all proposed by the Australian side, which laid down the essentials of how to run the project, and to design and to build the telescope. We comment on some of the clauses here; others are discussed in Chapter 4. A contract for the entire telescope, including both design and construction, now could not be let to any one manufacturer. This had been the usual British practice and was the course favoured by Woolley. Rather, a Project Office would be set up and the design carried out by the Project Manager and his staff, with the help of consultants. Again, tenders were to be invited on a world-wide basis. As Hoyle has emphasised, it is most unusual for a government allocating large sums for a purpose of this nature not to insist that most of it be spent in the home country. This condition could cost the British valuable contracts (and it did), whereas whatever else happened Australia had a virtual guarantee on the valuable building and dome contract.[13] In practice, major contracts went to Britain, Australia, the USA, Japan and Switzerland: much, we can safely assert, to the benefit of the telescope.

Another condition in the interim agreement was that the design should follow broadly that of the Kitt Peak telescope. There was no disagreement over this, but it was a political and not a technical recommendation; it is not mentioned in the submission made to the Governments in 1965. Possibly the British were determined that at all costs they must avoid the INT imbroglio; possibly the Australians had misgivings about British technology (the contract for their own Parkes telescope had gone to a German company); certainly many astronomers considered that as all the big telescopes were in America, only the Americans knew how to build them.

Finally, the AAT was the first large telescope to be built without a pre-existing infrastructure into which it could be incorporated. Clearly some form of administrative framework would be necessary, and the 1965 submission included the recommendations that 'an Australian–United Kingdom Astronomy Research Institute be set up by a convention between the two governments' and 'a headquarters should be established in a capital city' (see Appendix 2). However the matter is not referred to in the interim agreement. Had it been, a great deal of trouble might have been avoided.

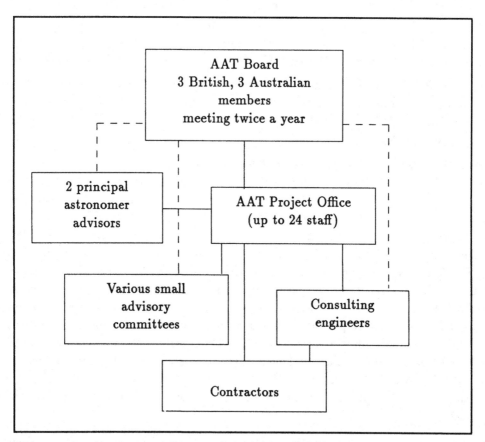

Figure 2.1   Organisation of design and construction of the AAT

2.5 The Project Office

The interim agreement laid down that overall responsibility for the telescope was to be vested in a body initially known as the Joint Policy Committee (JPC). It was composed of three members nominated by each Government, and it served until February 1971 when, following the passage of the Australian Act in 1970, it became the Anglo-Australian Telescope Board. The Project Office was directly responsible to these bodies, of which it acted as the executive arm as far as the telescope was concerned. However it was not practicable to set it up earlier than 1968, and in the meantime an *ad hoc* Technical Committee was invited to act. It consisted of Pope and Redman from the British side, and Gascoigne and Wehner from Australia. It reported to the JPC on all aspects of the telescope design – optics, mounting, drive and control, building and dome, site testing and site works – and pursued urgent matters like the acquisition of the primary mirror blank, and generally kept technical matters moving until a Project Manager was appointed.

The Technical Committee (without Pope) first met as such in Tucson on May 8 1967, where the AURA Board was meeting to finalise the design of the mounting for the Kitt Peak telescope. (J.F. Hosie from the JPC also was present.) The proposed design was duly adopted and put out to tender. Some 80 man-years of engineer, astronomer and draughtsman time had gone into it, and the associated mass of documentation was formidable. The meeting was the AAT group's introduction to the Kitt Peak mounting, with which it would become so familiar, and it also marked the beginning of a close and rewarding association with the Kitt Peak project staff. Following this, the Technical Committee was closely associated with the ordering of the primary mirror blank, with the first assessment of the Kitt Peak mounting, and with preliminary work on the site, building and dome.

The first Project Manager was appointed in August 1967, initially as Project Engineer, and took up office early in 1968. This was the beginning of the Project Office. Located in Canberra, the Project Office built up to a peak staff of about 24, mostly engineers and draughtsmen. It produced designs and specifications for the telescope and everything related to it (a great deal of this was original work), arranged for the calling and letting of tenders, supervised contracts and contractors, and managed the finances. A feature of its work, necessitated by the many innovations in the design, was the extensive use made of specialist consultants. Contact with the astronomers was maintained partly through Redman and Gascoigne, partly through various *ad hoc* committees, and by personal contacts.

In all, there were four Project Managers. The first was M.H. Jeffery. His was an excellent appointment, and his sudden death in September 1969 was a devastating blow. He had begun in the position early in 1968, and in his limited time had given the Project Office its early shape and direction, and had seen the outstanding structural problems of the telescope mounting satisfactorily resolved. Following his death, H.C. Minnett, an engineer-scientist from the Radiophysics Division, acted as Manager until June 1970. When he declined the offer of a more extended term, W.A. Goodsell from the British Ministry of Works was appointed for a three-year term. Goodsell returned to England to become Project Manager for the William Herschel Telescope, and was succeeded by H.P. Wehner, who managed the Office from June 1973 until it was wound up in July 1975.

Besides their managerial functions, each of these men made a characteristic personal contribution to the Project. Jeffery's we have mentioned. Minnett's was to the drive and control system, a field in which he had been expert since his days with the Parkes radio telescope, and he was largely responsible for the adoption of spur gears in the drive, one of the major innovations of the AAT. In Goodsell's term construction and the rate of expenditure of funds rose to a peak, developments in which his expertise in contractual matters had well fitted him to cope. When Wehner took over he had been with the Project longer and knew it better than anyone; he saw it through the delicate final stages and handed the completed telescope over to the Anglo-Australian Observatory.

At its meeting in August 1967 when the JPC appointed Jeffery, it engaged Pope and Wehner as full-time senior engineers; later in the year it made H.C. Minnett (CSIRO, Epping) and L.E. Ford (Royal Radar Establishment, Malvern) consultants to advise on drive and control. Of other senior people, D.W. Cunliffe (CSIRO, Melbourne) joined early in 1968 as Executive and Finance Officer, J.R. Rothwell (CSIRO, Parkes) as electrical engineer, and K. Hall was seconded from Freeman Fox and Partner to assist with the structural aspects of the mounting design. M. Beard (CSIRO, Epping) commenced in November 1971 as head of the computer group, while T. Wallace arrived in September 1971 on secondment from the Royal Observatory, Edinburgh, to take charge of the computer software. He died suddenly in November 1973, shortly before his contract ended. Finally, when John Pope returned to the Royal Greenwich Observatory in September 1971 he was succeeded in November by P.R. Gillingham, an Australian who had been at the RGO for several years. Gillingham and Cunliffe have remained with the AAO. This selection is to a certain extent arbitrary, and a full list of Project Office staff is given in Appendix 5.

We also mention the two consulting engineering firms: Freeman Fox and Partners of London, and Macdonald Wagner and Priddle of Sydney. Both had a considerable influence on the Project over an extended period. Four Freeman Fox engineers – M.H. Jeffery, K. Hall, C. Blackwell and D. Fern – were seconded to the Project at various times, and had much to do with the success of the mounting and drive, while D.H. Collins from Macdonald Wagner and Priddle played a similar role as site engineer for the erection of the building and dome.

It was obvious to all who knew him that the organisation and staffing of the

Project Office owed much to E.G. (Taffy) Bowen, especially to his experience with the Parkes radio telescope. The driving force from the outset behind that most successful instrument, he had been associated with every facet of it, including raising the funds, participating in all stages of the design, and organising the construction team. The stipulation that AAT tenders be called on a world-wide basis clearly stemmed from his experience at Parkes, as did the association between the AAT and Freeman Fox (the Parkes telescope was designed by Freeman Fox, and built by the German contractor, MAN). Bowen also brought into the Project Office a number of the old Parkes hands, among whom Jeffery, Minnett, Rothwell and Beard all distinguished themselves. In this context, Les Ford, who came from a similar background should also be remembered. The value of the experience brought from the radio world to the AAT has already been emphasised.

The Project Office was unusual in that it was headed by an engineer and not by an astronomer. Two astronomers were attached to the Office on a semi-permanent basis: R.O. Redman, Professor of Astrophysics and Director of the Observatories, at the University of Cambridge, and S.C.B. Gascoigne, Professor of Astronomy, at the Australian National University. They had no executive authority; for example, they could not authorise expenditure, sign contracts, or hire staff, but they were regarded as integral members of the Office, attended Board meetings and were expected to be familiar with all aspects of the telescope. For some aspects of the work, notably the optics, they undertook prime responsibility. This was a sensible arrangement which worked well. Superintending a project of this magnitude with its complex contractual, legal, financial and staffing arrangements is for engineers, not astronomers; but it was essential that the users, the astronomers, be represented on a continuing basis, and there were occasions, sometimes unexpected, when this representation resulted in vital contributions.

One memorable feature of the Project Office was its universal sense of purpose, never overtly expressed but none the less unmistakable. To many of the staff it was the most rewarding job they would ever have. It had captured their imaginations, they could put their hearts into it, and it was worth the best efforts they could produce. More than once the similarities have been noted between the building of a large telescope and one of the great medieval cathedrals. They share a nobility of purpose and a beauty of design, they issue continual challenges to the technology of their times, and they make almost daily demands on the craftsmanship and ingenuity of their builders. The comparison seems valid. As Fred Hoyle put it on a memorable occasion: *'The beauty and grandeur of a large telescope is well matched to the scientific purposes for which it is to be used. Those who step for the first time on to the main floor of the Anglo-Australian Telescope building will, I believe, become immediately aware of an unusual relationship between aesthetic quality and advanced technology. A large telescope is a good example of the things which our civilisation does well.'* [14]

Table 2.1.   *Breakdown of costs for construction of the AAT*[15]

	$A'000
Complete parts (mostly optical blanks)	879
Optical figuring and tube (Grubb Parsons)	1643
Mounting and drive	2574
Building and dome	4651
Consultations	1552
Project Office	2328
Computer	278
Instrumentation	1258
Site works	299
Other	470
Total Australian dollars	15,932

2.6 A summary of progress

In the first years the main effort went into design. The first real item of expenditure was the payment for the primary mirror blank late in 1969, and actual construction began when work started on the building early in 1971. Activity in 1967 centred largely on the procuring of the mirror blank, the optical design, without which little else could proceed, and an assessment of the KPNO mounting design. Much of 1968 was occupied with the redesign of the mounting, an investigation of a revolutionary proposal to use spur gears instead of worm gears for the main drives, and planning for the site works, building and dome. In 1969 it was decided, for the sake of better servo control, to stiffen the polar axis, and this also was a major design task, compounded with something of a crisis which centred on the declination bearings. Satisfactory solutions were found to all these problems, and the mounting contract was placed in August 1970.

Meanwhile work had proceeded on the design of the building and dome, and after delays which arose when even the lowest tender came out too high, the contracts were let late in 1970. 1971 and 1972 were peak construction years, with work proceeding at pitch on the telescope optics, the building and dome, and the mounting. Supervision of the relevant contracts made heavy demands on the Project Office. The design of the telescope drive was completed, and an extensive investigation made of the proposed computer system. Both of these necessitated major design studies by outside consultants. Work on the instrumentation also began, much of it undertaken by astronomical institutions in the two countries.

Most of the major components were delivered to site in 1973, and by the end of that year the telescope was largely assembled. Following the last stage, the installation of the unbelievably complicated electrical wiring system, it was handed over by the contractors in March 1974. The optics were installed and the first photographic plate taken on 27 April 1974. At an impressive ceremony the telescope was officially inaugurated by Prince Charles on 16 October 1974. However the task of commissioning – the identification and correction of innumerable faults and maladjustments, and the tuning of the telescope to yield its optimum performance – continued into the next year. The telescope finally went into general use on 28 June 1975. Highlights of this period were the implementation of a computer control system which was by far the most sophisticated and effective used up to that time, and the inauguration of a newly built digital spectrograph, a copy of one at the Lick Observatory, and the most efficient instrument of its kind then in existence.

The telescope had taken a little over eight years to build, and had cost almost $A16 million. The original estimates presented in April 1967 were seven years and $A11 million. The project had exceeded the time estimate by a year and the cost estimate by a factor of 1.455, corresponding to an inflation rate of five per cent, appreciably less than that in either Australia or Britain over the same period. In other words, after allowance for inflation, the construction cost was somewhat less than the estimate made eight years before: the financial control indeed had been close. Roughly three of the eight years were spent on design, three on construction, and two on erection and commissioning. A dozen major and many minor construction contracts and sub-contracts were let, and a similar number of design studies commissioned.

Though it is not clear from the figures in Table 2.1, the optics cost about $1.1 million, made up of $0.6 million for the blanks, from the first item, and $0.5 million for the figuring, from the second item. This is less than seven per cent of the total. The cost of the telescope was the sum of the first three items, $5.096 million, 32 per cent of the total, though this figure does not include designing and consulting costs. The dome and building cost 29 per cent, to which design and consultancy figures would again have to be added. We comment further on these figures in later chapters.

Notes to Chapter 2

1 Dreyer J.L.E. (1923). *History of the Royal Astronomical Society, 1820–1920* 212
2 Plaskett H.H. (1946). *Mon. Not. R. Astr. Soc.* **106**, 80 'Presidential Address'
3 Babcock H.W. (1978). ESO conference on Optical Telescopes of the Future, F. Pacini *et al.* (ed.) 37
4 For additional information on these matters see *The General History of Astronomy*, Volume 4, Cambridge University Press, 1984
   – Lankford J. 'The impact of photography on astronomy' pp.16–39
   – Van Helden A. 'Building large telescopes, 1900–1950' p.134
5 It was reckoned that the 200-inch mirror had the stiffness of a solid disc of the same

glass 14 inches thick, or as I.S. Bowen put it, the same as a disc 60-inch in diameter and 1.28 inches thick. The AAT mirror has a stiffness of a 60-inch disc 3.54 inches thick. The UKIRT mirror, with a diameter of 135 inches and a mean thickness of 9.6 inches, has the stiffness of a 60-inch disc 1.9 inches thick.

6 This statement did not necessarily hold for astronomers, as opposed to engineers. R.J. Weymann has related how a high-level committee (it later produced the Greenstein Report) was drawing up a priority list for future telescopes. 'At an early meeting in Pasadena [about 1970] a southern copy of the Hale 5-metre telescope was designated the top priority optical project.' Reaction at the meeting was such that 'the panel was persuaded to hear a presentation by Aden Meinel on the MMT concept'. The MMT (multiple-mirror telescope), then a very advanced idea, is described briefly in Chapter 14. See *Telescopes for the 1980s* ed. G. Burbidge and A. Hewitt, p.74

7 Preston R. *The New Yorker* 26 October 1987 p.81

8 The European Southern Observatory is a consortium originally of five countries which operates a large astronomical observatory in northern Chile. Established on 6 October 1962, its headquarters are in Munich. The original five countries – Belgium, France, The Federal Republic of Germany, The Netherlands and Sweden – have since been joined by Denmark, Italy and Switzerland.

9 Crawford D.L. (1965). *Sky and Telescope* **29**

10 cf. Hoyle F. (1982). *The Anglo-Australian Telescope*, University College Cardiff Press, p.6. The decision to adopt the alt-azimuth mounting for the William Herschel Telescope was made in 1969, on the recommendation of a special committee made up of R.O. Redman (Chairman), G.J. Carpenter, J.G. Davies, V.C. Reddish and T.A. Wyatt.

11 The submission is Annex 3 in Lovell (1985) *Qrt. J. R. Astr. Soc.* **26**, 393

12 Letter from Redman to Reddish 30 June 1967 (RGO 37/429)

13 Hoyle F. (1982). *The Anglo-Australian Telescope*, University College Cardiff Press, p.9

14 From his address at the inauguration of the telescope, 16 October 1974

15 AAT Annual Report 1976–77, p.34

3 The campaign for a large telescope

3.1 The political background

The years from mid-1950 until 1967 witnessed a tangled web of proposals, counter proposals, deliberations and negotiations for a large telescope in the southern hemisphere in both Australia and the United Kingdom, and to a lesser extent in the United States and in Europe. A detailed account of a British view of this period has been given by Sir Bernard Lovell.[1] However, before following the main strands of this web, it will be useful to outline the nature of the principal participants and the administrative structures within which each operated.

The organisational background to Australian astronomy was relatively simple in the nineteenth century: the state observatories were pre-eminent and their Directors were the main sources of scientific advice to the State Governors, but from the end of the century their relative importance was in decline. At this time, the drafting of the Commonwealth Constitution was under way, and after the 1897 Adelaide Convention Debate on the Constitution, the Founding Fathers saw fit to include among the powers of the new Commonwealth Parliament the power to make astronomical and meteorological observations.[2] This made possible the emergence of the Commonwealth Solar Observatory in 1924, later to become Mount Stromlo Observatory, and the Radio Research Board in 1926. Funding was not uniform; the Department of the Interior provided funds for the former, and the Council for Scientific and Industrial Research, the precursor of the Commonwealth Scientific and Industrial Research Organization (CSIRO), and the Post Master General's Department were the principal sources of funds for the latter.

Before 1954, scientific research in Australia required no formal body to assess and to fund scientific projects, since most were funded through the universities or the CSIRO which were the research centres. Some special projects might receive funds directly from the Government, in the way Woolley had successfully approached the Labor Prime Minister of the day, J.B. Chifley, for funds for Mount Stromlo's 74-inch telescope. However, by 1954 a prestigious body of eminent scientists had formed the Australian Academy of Science, which became the appropriate body best able to assess the claims of astronomy and to be responsible for supervising the preparation of submissions to the Government for funding of a large telescope project. Other avenues for obtaining funds could still be explored outside the Government; for example, E.G. Bowen, Chief of the CSIRO Division of Radiophysics, organised almost single handedly funds from the Carnegie Institution of Washington and the Rockerfeller Foundation in America[3] for the CSIRO's Parkes radio telescope which came into operation in 1961. In addition, the University of Sydney obtained over $US800,000 for its Mills Cross from the American National Science Foundation during this period.

A corresponding brief summary of the organisational background in Britain falls into two parts, with April 1965 marking the transition between them. Prior to 1965, astronomical research had long been carried out in a number of British universities, but the main thrust was through two Government observatories. One of these, the Royal Greenwich Observatory (RGO) whose Director was the Astronomer Royal, was among the oldest and best-known scientific centres in the world. It was funded directly and independently by the Admiralty, whereas universities received Government support through direct university funds, and additionally for larger projects through grants from the Department of Scientific and Industrial Research (DSIR). The premier scientific society in the country, the Royal Society of London, could obtain direct grants from the Treasury for projects of special merit. The Isaac Newton Telescope was authorised as a result of Royal Society sponsorship, reinforced by that of the Royal Astronomical Society. On completion it was installed at the RGO. On the other hand, the big radio telescopes at the Universities of Cambridge and Manchester were funded largely by grants from DSIR. The other Government observatory was the Royal Observatory, Edinburgh, whose Director was both head of the Astronomy Department of Edinburgh University and Astronomer Royal for Scotland. The interests of the two Government observatories were mainly in optical astronomy and related instrumentation.

To obtain a wider and independent view of the merits of competing claims for the support of scientific projects, the Government relied on the advice of the Advisory Council for Scientific Policy, but this body was not equipped for the detailed assessments increasingly necessary as projects became ever more complex and expensive. It was clear that a more coherent framework for the assessment of funding of large projects was needed, and after a thorough review by a committee under the Chairmanship of Sir Burke Trend, the Government introduced fundamental changes in April 1965 when the Science Research Council (SRC) was one of several new statutory councils established by Royal Charter. The SRC, with a full-time Chairman and membership drawn from experienced academic scientists and industrialists, was given far-reaching powers to assess and to provide financial support to projects over a wide range of the physical and biological sciences. Applicants for funds for major 'big science' such as nuclear physics, radio astronomy and space research could no longer seek financial support through differing channels, thereby placing the onus of making decisions between them on organisations ill equipped for the task. All would now be assessed and decisions taken within one appropriate organisation, the SRC. The SRC operated under the aegis of the Department of Education and Science; it was autonomous in its scientific judgments, but subject to governmental control in matters having major policy or financial implications.

The SRC was one of the new statutory councils in 1965; a concurrent creation was the Council for Scientific Policy. The latter's responsibilities included the recommended allocation of a science budget among these councils, without becoming involved in the project by project deliberations. The Chairman was Sir Harrie Massey, Professor of Physics in University College London; an Australian by birth, he was a strong supporter of international co-operation in scientific projects. His familiarity with the 150-inch project as Chairman of this council was to add weight to his discussions with the Australian Minister in 1966.

The effect of this change on the negotiations concerning southern hemisphere telescopes was evident mainly in the reduced powers of the Royal Society and of the Royal Greenwich Observatory to act independently. Control of the RGO had passed from the Admiralty to the Science Research Council; the Royal Society could exert influence but no longer could submit claims for significant funds for astronomy. The SRC became the negotiating agency, and within it, the opinions from all branches of astronomy could be heard, as could opinions from other disciplines competing for the limited funds available.

3.2 A large telescope project

The suggestion that there should be a large optical telescope in the southern hemisphere associated with the British Commonwealth seems to have come first from Woolley in 1953 while Director of Mount Stromlo Observatory. Indeed, Professor Sir Mark Oliphant, Director of the ANU's School of Research in Physical Sciences, wrote to Sir John Cockcroft, a famous British physicist, in 1954 outlining a plan he and Woolley had 'cooked up' to build a very large telescope at Mount Stromlo.[4] Woolley contemplated a 200-inch instrument as a collaborative project between Australia, Britain and Canada, to be located in Australia. The idea received strong support from B.J Bok, Woolley's successor at Mount Stromlo Observatory. Canadian interest did not develop, but the seeds had been sown in the astronomical communities of both Australia and Britain where Bok and Woolley respectively promoted with energy and enthusiasm the general idea of a southern hemisphere telescope. There were, however, complications which could not be ignored. British astronomers already had access to telescopes in South Africa at the Royal Observatory at the Cape, and at the Radcliffe Observatory in Pretoria, and, while there might have been attractions in building on these connections, the political outlook for the future of long-term scientific investments was not promising. In Western Europe there was much preparatory work in progress by astronomers in Belgium, France, the Federal Republic of Germany, The Netherlands and Sweden with the aim of establishing a European Southern Observatory (ESO) in Chile, and British participation was actively sought. Some Australian and British astronomers could see many attractions in collaborating with astronomers from the USA. The Americans had played a leading role in optical astronomy since early in the century, and several American groups were striving to establish big telescopes in the southern hemisphere in the late 1950s.

The conceivable collaborations were numerous and most had political overtones, particularly those concerning Britain and Western Europe at a time when the relations of Britain with the European Community were being widely debated.

Woolley, supported by many British astronomers, vigorously opposed British membership of ESO mainly on the grounds that sharing the available telescope time with so many partners would give Britain too little for its needs and in return for its financial contribution. It also effectively would quash any chance of a British Commonwealth project.

In 1959 the British National Committee for Astronomy, one of the Royal Society's committees, considered future plans for optical astronomy, and recommended that first priority be given to developing the Radcliffe Observatory in South Africa as the initial step towards improving observing facilities in the southern hemisphere. For the longer term, it proposed a British Commonwealth Observatory under joint management in Australia. The views of this Committee were supported by the Council of the Royal Society and passed to the Advisory Council for Scientific Policy in late 1959. This was an important formal step towards bringing the needs of British scientists in southern hemisphere astronomy to the attention of the British Government, but there was still a long way to go. Although astronomers were in little doubt about their attitudes to ESO, politicians appeared to be keeping options open, to the extent that the Australian Prime Minister, R.G. Menzies, on a visit to London in 1959, was said to believe a British Commonwealth project was unlikely to find favour over a European collaboration.

The President of the Royal Society attempted to revive interest in a British and Australian project when he made the first formal approach to the Australian Academy of Science in late 1959. When the matter was mentioned to members of the Academy it brought a surprising result. Seven Fellows eminent in the biological sciences expressed fears that such a 'gigantic' expenditure would adversely affect funds for research in other areas. The debate was well-known and at times heated with one irate biologist declaring that the astronomers wanted only 'to keep up with the astronomical Joneses'. In March the following year, Bart Bok and Sir Mark Oliphant addressed the Council of the Academy on the telescope proposal, and the Council then endorsed the project scientifically but asked for more time and further information to study the financial and administrative implications. This endorsement led directly to the setting up of a Fauna and Flora Committee to advise the Council on major biological projects.

The two years to mid-1960 witnessed no progress towards a Commonwealth collaboration. Woolley tried unsuccessfully in Britain to gain support for a British, Canadian and Australian project, and maintained his strong opposition to British membership of ESO. The British Government eventually accepted his advice against joining ESO in 1962. In Australia the Government asked the Academy in June 1961 to consider the merits of Australia's joining a British Commonwealth Observatory. However, while supporting the scientific merit of such a proposal, the Academy still wanted to investigate the various collaborative options. At the same time, it considered other possible scientific projects among which those in the biological sciences were conspicuous. It proposed reporting to the Government within two years on all 12 proposals before it, ranked in order of priority. It was clear to the British that there was insufficient general support for a British Commonwealth project, and to the Australians that there could be a two-year delay on any large telescope project.

Following the reaction of the Academy, Bart Bok spoke out publicly in August 1962 against such a delay in his Presidential Address to Section A of the Australian and New Zealand Association for the Advancement of Science (ANZAAS).[5] Bok gave credit to Woolley and Oliphant for having made the first suggestion as early as 1955 that Australia should become a major partner in a British Commonwealth Southern Observatory. In evaluating preliminary proposals, the Council of the Australian Academy of Science was the principal channel for advice to the Government. Bok's criticism of the Academy was that the Council did not consult any Australian optical astronomer or the country's younger astronomers through its own National Committee on Astronomy when preparing a report for the Government in 1961. The written version of Bok's ANZAAS address does not include his criticism in the strongest terms that the Academy was in fact slowing things down. He accused the Academy of failing to show sufficient awareness of the need for a positive reply when the first enquiries came from the President of the Royal Society, Sir Cyril Hinshelwood, in 1959. He regarded the Academy's delay as prompting the British to seek partners elsewhere because, for the latter, mid-1962 was probably the nadir of a potential collaboration between them and the Australians.

Not surprisingly, the response of the then President of the Academy, T.M. Cherry, was sharp. He accused Bok of having offended the Establishment, hence having outlived his usefulness to Australia, and said that it was time he returned to the United States. Bok understood the Establishment to mean the Prime Minister, R.G. Menzies, with whom he had enjoyed good relations. Sometime later Bok and Menzies were present at an official function at the Australian National University. Without moving his eyes from the official procession, Menzies walked past Bok and delivered him a punch in the stomach and an eloquent message, *'You are a bad, bad boy!'*[6] Bok realised that he had *not* offended the Establishment, at least not where it counted.

In defence of the Academy during this period, it was less than ten years old in 1963. It did not enjoy the power, prestige and access to funds to anywhere near the extent that the Royal Society did in Britain, where it had some 300 years of experience in influencing government. Two years later Cherry wrote to Harley Wood, Chairman of the Academy's National Committee of Astronomy:

The Academy is well aware of the astronomers' case, and is sympathetic to it; but it is also aware of desirable projects in other branches of science. In my view Australia can afford to pursue the lot, without significant scientific competition. I just don't know whether that will be the government's view; between ourselves, the difficulty over these past years has been to discover whether the government has any coherent view.[7]

The situation began to improve as a result of talks in Australia between Sir Harrie Massey and senior political and scientific figures, when Massey reported in September 1962 that he had found much Australian interest in the possible telescope project. It was confirmed in the following January that at least some of this interest extended to Australian Government circles, when Bok and Hermann Bondi, then Secretary of the Royal Astronomical Society, discussed the matter with the Australian Prime Minister's Department. Exploratory talks between scientists of both countries followed.

The first important meeting took place in Australia in March 1963. The participants were R.v.d.R. Woolley, Sir John Cockcroft and L.G.H. Huxley (Chancellor and Vice-Chancellor respectively of the ANU), T.M. Cherry (President of the Academy), Sir Mark Oliphant, B.J. Bok, S.C.B. Gascoigne, E.G. Bowen and B.Y. Mills. The outcome was unanimous agreement that a strong case existed for a large optical telescope in Australia, and that steps be taken to encourage the British Government to make a proposal to the Australian Government for a joint project. This positive result from a group of eminent and influential scientists gave a new impetus to discussions. In the same month, Menzies, the Australian Prime Minister had introduced in the Parliament the Australian National University Bill, which proposed legislation to permit the University to establish a field station outside the Australian Capital Territory. Debating of this bill provided the much-needed opportunity to air publicly the case for a large telescope project during April and May, with the result that it received prominence and support. Further attention was drawn to the proposal with the publication of Gascoigne's timely article *Towards a Southern Commonwealth Observatory* in March.[8]

In Britain, it was Woolley who put great personal effort into the project.[9] He convinced the British National Committee for Astronomy and, through it, the Council of the Royal Society and the Advisory Council for Scientific Policy of the scientific case and the need for speedy action if important opportunities were not to be lost. In May, Woolley wrote to Bok and proposed setting up a small working party, confined to astronomers who used large telescopes: R.v.d.R. Woolley, R.O. Redman, A. Hunter and W.L.W. Sargent from Britain; S.C.B. Gascoigne, A.W. Rodgers and B. Westerlund from Australia; I.S. Bowen from the USA. It met at Herstmonceux in August to consider design specifications for a possible southern hemisphere telescope. However, Australian astronomers were apprehensive about attending as individuals at British expense for fear of being seen as junior partners. Consequently, Oliphant initiated a plan for the Australians to attend as representatives of the Academy. In addition, he suggested that the Academy establish a committee to deal with questions of justification, telescope design and Anglo-Australian co-operation. This committee would have the power to nominate representatives to discussions in Britain or the USA and to approve all technical proposals, but without necessarily committing the Academy or the Government to any particular course of action. Accordingly, the Academy set up a Committee on Proposals for a Large Optical Telescope under the Chairmanship of Huxley, and comprising B.J. Bok, W.N. Christiansen, S.C.B. Gascoigne, B.Y. Mills, Sir Mark Oliphant and Harley Wood. During a visit to Australia, the President of the Royal Society, Sir Howard Florey, was informed of this new committee and was invited to set up a complementary Royal Society committee. A joint British committee of DSIR and the Royal Society was soon in place comprising R.v.d.R. Woolley, D.E. Blackwell, H. Bondi, W.H. McCrea, J.D. McGee, Sir Harrie Massey, R.O. Redman, M. Ryle, Sir Harry Melville, and with Florey as Chairman. It first convened in December 1963.

In the meantime, the Herstmonceux meeting of astronomers had considered in some detail technical specifications of the proposed telescope. There was agreement that the telescope should have an aperture of 150 inches and an equatorial mounting. Bok reported that the authorities in the USA had offered to make available the design studies for a similar telescope to be erected at Kitt Peak, in Arizona; he also advised that there would be strong pressure that the tendering procedures for the telescope design and construction should follow the Australian practice, a reference to the Australian dislike for the British practice of placing an overall contract with one main contractor. These two statements by Bok were to have a major influence on the eventual project.

The Australian Committee on the Large Telescope acted swiftly and by February 1964 could submit its final report to the Council of the Academy. In June a scientific mission from Britain visited Australia, and drew up the case for the 150-inch telescope which included an outline of the specification for the instrument with the technical sub-committee of J.C. Bolton, S.C.B. Gascoigne, A.W. Rodgers and H.P. Wehner. There was agreement on key parameters in the specification, and a conclusion that a site on Siding Spring Mountain was as good as any in Australia on which information was available, although testing of two other sites, at Mount Serle in the Flinders Ranges and Mount Singleton in Western Australia, was to continue. At this stage the estimated capital cost was £3.57 million if located at Siding Spring, with annual running costs of £126,000. After the Academy and the Royal Society considered reports from their respective groups, it was agreed in June 1965 that simultaneous approaches would be made to their appropriate Governments on the basis of a joint equal British-Australian project. In Australia the submission was presented to Senator John Gorton, Minister in Charge of Commonwealth Activities in Education and Research. In Britain, the Secretary of State for Education and Science, the Right Honourable Anthony Crossland received the report. The capital cost was now estimated to be £4 million spread over eight years, with £440,000 for planning and site testing. Running costs were estimated to be £146,700 per year.

It was at this stage of the developments, as described earlier in this chapter, that the British Science Research Council had newly come into being, and the assessment of priorities and control of the relevant funds passed into the hands of the Council. The Royal Society was no longer able to take direct action, but was still in a position to exert an influence in various ways. The new Science Research Council was faced with substantial claims for the support of new projects, not only in astronomy, but also in other expensive fields of 'big science', particularly in space research and nuclear physics.

For the first time these claims were in direct competition; in the case of astronomy, there were proposals for new radio telescopes at both the Universities of Cambridge and Manchester, for an Institute of Theoretical Astronomy at Cambridge, for new telescopes proposed by the two Royal Observatories and for the proposed 150-inch telescope in Australia. It was soon clear that the likely allocation of funds for astronomy would very severely limit the number of projects which could be approved, and much hard discussion was necessary before the Astronomy Space and Radio (ASR) Board of the SRC advised Council that of all the proposed new projects in astronomy, the highest priority should be given to the Anglo-Australian Telescope. The Council, concerned at the heavy commitments for which the ASR Board was already responsible, at first was not prepared to agree even to the modest expenditure required for the initial design studies for the AAT. However, the arguments put forward particularly by Sir Bernard Lovell, Chairman of the ASR Board, and Hermann Bondi finally convinced Council of the strength of the case. Lovell, who wanted money for radio astronomy at Jodrell Bank, conceded that there was a greater need for optical telescopes at the time. It was agreed to recommend to the British Government to open discussions with the Australian Government on the basis that the British would be prepared to meet half the cost of site investigations and the detailed design study which would be an essential preliminary to the full-scale project.

The formal proposal to the British Department of Education and Science was made in February 1966, and after due consideration the Department with Treasury approval authorised a formal approach to the Australian Government. In June an *aide memoire* informed the Australian Government that the British Government was willing to share the cost of the design study and other preliminary actions, and proposed that the two Governments should establish a Joint Preparatory Commission to oversee the work. The reply from the Australian Prime Minister's Department in July was disappointing to the many British astronomers who had recommended such high priority for the 150-inch project. This reply indicated that the Australian

Government was expecting an offer of a joint project with the USA. For this reason no decision to proceed with a project had yet been made, and there was no case for establishing a Joint Preparatory Commission.

In Australia towards the end of 1965, Senator John Gorton, the Minister in Charge of Commonwealth Activities in Education and Research, soon to become Minister for Education and Science, was eager for a large scientific project which his new Department could sponsor. Among the proposals before him were those for a large optical telescope. He had a strong preference for the 150-inch proposal, and as a good nationalist, ideally as a completely Australian project. This sense of nationalism probably accounts for his original support for it as a 200-inch telescope to rival the Palomar 200-inch. However, he knew that his senior colleagues in the Cabinet and the Prime Minister would not find funds for a project of this magnitude. Falling back on the second option, a partnership, he was aware of Menzies' disenchantment with the British as a possible partner. Therefore, before presenting a proposition to the Cabinet, Gorton wanted to show that he had explored options for potential partners. In addition, he wanted to place two separate proposals before Cabinet from which it could choose the more favourable. From the possible partnerships, his preference was for collaboration with the Americans.

In November, Gorton met with Huxley and Bowen, both of whom were strong supporters of the 150-inch proposal. Recognising Bowen's familiarity with the scientific community in the USA, Gorton asked him to visit his American colleagues and explore whether any group would be willing to join a 150-inch telescope project in Australia on a fifty-fifty basis. Bowen's approach to American scientific and financial groups in January 1966 came too late because plans for their own telescopes already were well advanced. The astronomical groups consulted were the Carnegie Institution of Washington, California Institute of Technology, the Kitt Peak National Observatory and the University of California (Berkeley, Santa Cruz, UCLA and San Diego). Of these, the Lick Obseratory of the University of California was the only potential partner, and later in 1966 it sent its Director, Albert Whitford, to Australia to investigate sites. Such was the degree of interest in an Australian collaboration, but Lick Observatory's reply to the Australian Government was slow. The British were aware of the approaches which Australia was making to the Americans and of the Australian preference for a connection with a group in the United States. As long as the possibility of a positive response from the University of California remained open, the Australian Government replied to the British with the *aide memoire* of June 1966 in unenthusiastic terms.

The pace of events was painstakingly slow to astronomers in both Britain and Australia. Although the SRC had resolved the conflict of priorities in favour of collaboration with Australia, serious, if unofficial, discussions continued with leading American astronomers about possible co-operative projects. In particular, there were talks between Woolley and Horace Babcock, Director of Mount Wilson and Palomar Observatories, in the USA in June 1966. From these discussions came a proposal for an Anglo-American project for three telescopes in Chile, which reached the stage of a draft Memorandum of Agreement, but which presented serious funding problems.

British astronomers continued to press the seriousness of their Government's offer to collaborate with Australia, and on one of his private visits to Australia in July, Massey met Gorton. He assured Gorton that the proposed Joint Preparatory Commission was *not* procrastination, but rather an indication of firm commitment, subject only to reaching satisfactory conclusions on various estimates of costs, time scale and so on, all of which would be examined by the Commission. Other assurances were forthcoming, and in September the Australian Government made an independent enquiry in Britain about funds. Sir Frederick White, the Chairman of CSIRO who was in London at the time, was asked to make a direct approach to the SRC on the status of the proposal. He was assured that it had the full support of the Council, which had included suitable financial provisions in its forward plans.

The situation was further clarified during a visit by W.L. Francis, the SRC Secretary, to Australia in October. He confirmed the assurances already given to White, explaining that the Council had a guarantee of funds which would enable it to start the project and an expectation of funds subsequently which would enable the Council to complete its share of the project. Accounts of these discussions, with varying degrees of accuracy, appeared in the Australian press on 25 October. Senator Gorton was quoted as saying that no binding decision had been made by either Government, but that the telescope, if built, would probably be financed jointly by the United Kingdom and Australia, and probably sited at Siding Spring.

Meanwhile the prospects of an Australian-American collaboration diminished, and by October it was apparent that such a partnership was most unlikely.

In Australia, in the early months of 1967, Huxley and Bowen had lobbied Senator Gorton and found that he was very much in favour of the 150-inch telescope. However, continuing uncertainty in British astronomical circles about the Australian Government's intentions caused contingency plans to be brought forward in case the long hoped for Anglo-Australian project had to be abandoned. The Astronomy Policy and Grants Committee (APGC) re-examined the programme for optical astronomy. It re-affirmed the correctness of not joining ESO, and the priority given to the proposed AAT should the Australians agree. Failing Australian agreement, it foresaw a possible collaboration with American groups in Chile, which Woolley had explored with Babcock, as a way of obtaining improved access to southern skies. However, the APGC decided the establishment of a new large telescope in the northern hemisphere should receive priority if the AAT did not proceed. H.A. Brück, Astronomer Royal for Scotland, had long advocated such a project, and a substantial proposal for such a telescope in the Mediterranean region was advanced by three Cambridge Professors, Hoyle, Redman and Ryle, and had received much support. It was not until the mid-1970s that the northern hemisphere observatory in the Canary Islands developed as a joint British and European project.

Out of this myriad of possibilities which presented itself to the astronomical communities of Britain and Australia, out of the hundreds of hours of official and informal communications between groups and governments on three continents, and out of lengthy debate among scientists for funds for competing projects – all of which characterised events in the 1960s – was born the Anglo-Australian Telescope project. On 14 April 1967, more than ten years after Woolley first spoke of a large telescope in Australia, the British Government received Australia's formal agreement to open substantive negotiations. The Australian response was, in effect, a new proposal and included a set of conditions, in particular, that the Kitt Peak design would be used, which demonstrated an unwillingness to become involved in a protracted design study and preparatory work. So drawn out had been the period of negotiation, that most members of the astronomical communities in both Australia and Britain were no less amazed as delighted at the result. Redman expressed the sentiments at the time in a letter to C.S. Beals, his brother-in-law and the Chief Government Astronomer in Canada:

As you have heard, the AAT has suddenly come to life, after most people in this country had given it up for dead and, instead, during recent months all kinds of ambitious alternatives have been discussed here.[10]

A Joint Policy Committee was quickly established, which met in London in August, thereby ending a long period of uncertainty for astronomers of both countries, and opening the way for real development of the project.

Notes to Chapter 3

1 Lovell A.C.B. (1985). *Qrt. J. R. Astr. Soc.* **26**, 393
2 Constitution of the Australian Commonwealth – Part V, Section 51(viii)
3 The Menzies Government matched these grants pound for pound
4 Cockburn S. and Ellyard D. (1981). *Oliphant: the Life and Times of Sir Mark Oliphant* Axiom Book, p.178
5 Printed version in *The Australian Journal of Science* January 1963, p.281
6 Bok's own account undated, but written in the 1960s
7 Letter Cherry to Harley Wood, 23 January 1965
8 Gascoigne S.C.B. (1963). 'Towards a Southern Commonwealth Observatory' *Nature* (London) **197**, 1240
9 See the biographical memoir of Woolley by McCrea W.H. (1989) *Historical Records of Australian Academy of Science* **7**, 315
10 Letter Redman to Beals, 4 May 1967 (RGO 37/299)

4 An Anglo-Australian agreement

4.1 The negotiations

The message which signalled the start of the large telescope project in earnest came from Australia on 14 April 1967, following discussions over many years described in the previous chapter. In essence, the message said that the Australian Government was prepared to join the British Government as equal partners in building and operating a large optical telescope in Australia. The estimated capital cost in 1967 prices was $11 million (£4.4 million) spread over six years, with $450,000 annual operating costs thereafter. Gascoigne and Wehner made these estimates based on the Kitt Peak telescope project. The conditions which the Australian Government stated were that the telescope should be built on Siding Spring Mountain, in New South Wales; that the design should follow that of the Kitt Peak 4-metre telescope in USA; that there should be a sharing of facilities with the Australian National University which had established an observatory at Siding Spring; that there should be a joint controlling body with equal British and Australian representation; that observing time be shared equally between the two partners and control of the operation of the telescope be the responsibility of a person chosen by the controlling body; and that there should be supervision of the construction, following world-wide tenders and both Governments' approval of the major contractors.

Absent from the conditions was any reference to establishing an Australian–United Kingdom astronomy research institute which would be the body to manage the affairs of, and to operate the telescope. Both the Royal Society and the Australian Academy of Science made such a recommendation to their respective Governments in 1965; the submission is reproduced in Appendix 2. That there was this omission in the interim agreement caused a great deal of trouble between the Joint Policy Committee and then the Board, on the one hand, with the Australian National University. Redman foresaw the problems and often pressed the point that they could and should have been avoided if the Royal Society and Academy's explicit proposals for the administrative structure had been adopted.

Ten days after receiving the official message, a meeting took place in London between K.N. Jones and O.J. Eggen from Australia, and W.L. Francis, H. Bondi and J.F. Hosie from Britain. They were later joined by R.O. Redman, Sir Martin Ryle and Sir Richard Woolley. It was agreed that the Australian conditions would be acceptable to British astronomers, and that the way was now clear to establish working arrangements to implement the project. It was proposed to establish a Joint Policy Committee (JPC) to operate until a permanent structure formally could come into operation. In particular, the JPC would prepare for and be responsible for starting

Figure 4.1   JPC and Board member (1967–72) Sir Richard Woolley, OBE, FAA, FRS
(Photo: Australian National University)

the construction phase; it would propose the draft agreements to define the legal
framework within which the project would formally operate, and to allow allocation
of funds by the two Governments. Many other tasks would fall to the JPC, including
the finalising of specifications and the letting of construction contracts.

The Science Research Council (SRC) in Britain and the Department of Education
and Science (DES) in Australia were the operating agencies of each Government
responsible for the project. These two agencies were by no means mirror-images of
one another. The Australian Department had direct access to its Minister, whereas
the SRC was one step removed from its Minister with the British Department of
Education and Science existing in between them. A feature of this structure in
Britain's favour was that the SRC, through its system of scientific committees, was
always very well appraised of the needs and wishes of its astronomical community.
However, the same could not be said of government representatives in Australia. This
apparent lack of administrative symmetry assumed importance during the troubles
in 1970–73 over the management of the telescope, described in Chapter 8. Although
the Australian side was in a much stronger political position, the British agency was
clearly in a healthy position to argue the scientific case of the British community.

The two agencies had to agree to detailed financial arrangements. Although the
Treasury regulations and methods in both countries were broadly similar, they were
not identical, and the differences caused a number of problems which had to be
resolved. The SRC was required to obtain the agreement from its Department of
Education and Science in all major matters of policy and finance involved in estab-

lishing an international organisation such as that envisaged for the new telescope. This Department was in turn required to obtain Treasury approval before financial commitments could be made in advance of the formal inter-governmental agreement.

The JPC expected that the item most likely to cause delays in the construction phase of the telescope was the primary mirror, and that the blank should be ordered and obtained, and the slow process of figuring the mirror commence at the earliest possible moment. The British Department had given authority to the SRC in early May 1967 to proceed with the detailed planning of the Anglo-Australian Telescope project, and the SRC soon sought authority to proceed with tender action for the supply of the primary mirror blank. Towards the end of July, it had approval only for inviting tenders, but not for commitment to purchase until the inter-governmental agreement was concluded and the formal administrative structure was in place.

At this time the agencies in both countries were considering the various alternative formal arrangements for the project. Since the JPC was no more than an embryo governing body, a legal entity with full legal status under Australian law would have to replace it. In Australia the formation of a statutory authority by an act of parliament was the preferred option, although the British feared that this system would suffer from inflexibility and the long time scales of parliamentary operations. The SRC favoured either a registered company or a body incorporated by Royal Charter, but these were not commonly used in Australia for the type of operation in question. Since the administrative entity had to operate under Australian law and in accordance with Australian financial regulations, the decision on the appropriate legal structure was largely one for the Australian agency.

Pressure was increasing for action on several matters, and in August, Senator John Gorton, the Australian Minister for Education and Science, proposed to his opposite number in Britain, Anthony Crossland, that the appropriate agency, either Australian or British, should be authorised to engage consultants and staff, and in particular to purchase a mirror blank in accordance with the requirements of the JPC before completion of the inter-governmental agreement. In September, the British Secretary of State, who was by then R.St.J. Walker, agreed to these interim arrangements subject to them lasting for as short a time as possible.

Meanwhile, the JPC held its first meeting in London in August 1967. The membership for Britain was Sir Richard Woolley, H. Bondi and J.F. Hosie. Woolley was a brilliant and many-talented man of great presence and style, equally at home in scientific, academic and governmental circles. He was the Director of the Royal Greenwich Observatory and Astronomer Royal, and had been Director of Mount Stromlo Observatory between 1939 and 1955. The concept of the AAT was essentially his, and his early advocacy was most important in that it kept the project alive during a crucial period.

Figure 4.2   JPC and Board member (1967–73) E.G. (Taffy) Bowen, CBE, FAA *(Photo: CSIRO Division of Radiophysics)*

As it happened, Bondi was very soon appointed Director-General of the European Space Research Organisation (the precursor of the European Space Agency) and made only a brief appearance at the first JPC meeting. He was replaced at the next meeting by Fred Hoyle, Plumian Professor of Astronomy and Experimental Philosophy at Cambridge University, who was to have a long association with the project covering some of its most difficult days. Of the earlier members, he was the longest to serve the JPC and the Board, and on retirement more than seven years later wrote a personal account of his time with the AAT.[1]

The third British member, Jim Hosie, was the Director of the Astronomy, Space and Radio Division of the SRC. He had studied at Cambridge with Fred Hoyle, and gained his early training in the Indian Civil Service.

For Australia the membership was E.G. Bowen, O.J. Eggen and K.N. Jones. Bowen was the only member of the JPC and later the Board who had previous experience with the construction of a large scientific instrument. From 1946 to 1971 he had been Chief of the CSIRO Division of Radiophysics, during which time he was responsible for the construction of the 210-foot radio telescope at Parkes, New South Wales, and his experience in the many subtle problems associated with international tendering and contractors was invaluable.[2] In addition, as a result of his wartime involvement with the development of radar in Britain, he belonged to that generation of scientists who had won the respect and confidence of members in the top echelons of government. His would be a leading role in the success of the telescope.

Figure 4.3   JPC and Board member (1967–73) O.J. Eggen *(Photo: Australian National University)*

Olin Eggen was the Director of Mount Stromlo and Siding Spring Observatories from 1966 to 1977. He is an astronomer of the old school whose life and life-style have been determined wholly by astronomy. His record for time spent as an observer stands high internationally. His strong personality, forthright style and refusal to compromise with what he sees as the best interests of his subject have earned him powerful friends and equally powerful enemies. Many associate his name with the years of conflict between the Board and the Australian National University over the independent management of the AAT.

The Australian Government representative was Ken Jones, First Assistant Secretary in the Department of Education and Science. Often it was he who calmed the troubled waters which divided the Board and the University during the many difficult debates over the management.

Despite the differences in personalities of those who served on the JPC and the Board, there was one feature common to them all. They saw the AAT project as an opportunity to create a first-class scientific facility, and every one of them wanted it to excel internationally.

The first tasks of the JPC were to establish detailed interim financial arrangements, including the levels of delegated financial authority. It decided that the project office should be in Canberra, that the Australian DES should be responsible for

Figure 4.4   JPC and Board member (1967–73) K.N. Jones, CBE *(Photo: Ken Jones)*

the management of accounts, and that salaries and conditions of service should be based on those in the Australian Public Service. The interim arrangements were accepted generally by the British Government, but one matter of detailed difference between the practices in the two countries took considerable time to resolve. This concerned the currency in which transactions should be specified and whether or not adjustments should be made for exchange rate fluctuations. The JPC debated at length whether contracts should contain a clause allowing for price adjustments in the event of such fluctuations, since devaluation clauses were common in Australia, but strictly against British practice, even in the case of jointly placed contracts. The Australian members wanted all contracts to be placed from Australia under local procedures, and the matter was only resolved in April 1968 when the SRC sought British Government authority to invite tenders and to place further contracts relating to the mirror systems. The result was that these and later contracts were placed from Australia, with no devaluation clause, and prices were quoted in the currency of the country of manufacture. The first contract to be placed was with Owens-Illinois Inc. for the primary mirror blank in November 1967, and this was done from Britain according to British procedures and in pounds sterling. It was a favourable beginning to the project that the later devaluation of the pound was to the advantage of the JPC under this contract.

In March 1968, the JPC was able to make the important staff appointment of M.H. Jeffery from Freeman Fox and Partners, London, as the first Project Manager. Already the interim body had made significant progress negotiating contracts, engaging staff, and preparing the formal agreement between the Governments. In parallel with this, the Technical Committee had the scientific, technical and engineering planning well under way. The first Australian draft of the formal agreement came to Hosie in February 1968. It was clear that there would be an inter-governmental agreement and an act of the Australian Parliament, to which the agreement would be annexed, creating the Board as an Australian statutory authority. The Act, to be called the Anglo-Australian Telescope Agreement Act, would be administered by the Minister for Education and Science, through his Department.

Negotiations seemed slow and continued for most of 1968 and 1969, but during this time various sections within the Australian Government, as well as various Government Departments in Britain, had to examine both the Agreement and the Act. While the detailed legal aspects were in the hands of the Attorney-General, it was necessary to consult the Department of External Affairs because the Agreement, which would be concluded at the highest level for an international contract, namely with treaty status, impinged on Australia's foreign relations. In addition, the Treasury was concerned with the detailed financial ramifications of the project. An amendment to the Sales Tax legislation was required in anticipation of the AAT Agreement Act to avoid liability by the Board for the payment of sales tax on goods owned by it and for use associated with the telescope. Both the Treasury and the department concerned with customs and excise were consulted in this regard. The Act covered matters which either were not possible to be dealt with in the Agreement or were more correctly the subject of Australian domestic law. This included such things as the legal personality of the Board itself, Australian membership of the Board, taxation concessions, rights of employment of staff, and authority for the preparation and audit of accounts. At the same time detailed financial regulations were agreed under which the Board would operate on a daily basis.

4.2 The agreement
The Anglo-Australian Telescope Agreement was signed on 25 September 1969 in Canberra by the Australian Minister for Science and Education, Malcolm Fraser, and the British High Commissioner. It would come into force when the Australian Government notified the British Government that legislation to establish the Anglo-Australian Telescope Board had been enacted by the Australian Parliament and proclaimed. There remained a few points of difference to be settled in the wording of the financial rules, most of which arose from the differing practices in the two countries, and concerned matters such as the levels of delegated authority and the procedure for auditing accounts. Hosie rebutted the charges by the British DES that the Agreement should not have been signed with differences still to be settled by pointing out that there was pressure to complete the Agreement and consolidate the general position, that the points of difference were relatively minor, and that negotiators must be given discretion in such matters. When all seemed settled in May 1970, Jones raised an objection to a rule which would have allowed either agency,

properly authorised to incur substantial expenditure on behalf of the Board, to deduct such expenditure from its quarterly subscription to the Board's funds. His objection arose from the Australian Auditor-General's requirement to examine all supporting documents and not merely to rely in some cases on the British Auditor's certificates. The implication ran counter to general British practice which would require advances by the SRC to be kept to the minimum, and not made before money was actually needed. The Australian view, on the other hand, was that the Board must receive its contributions in full from each party, and keep its accounts in respect of gross receipts. There was eventual agreement on the basis that the Board would practise gross accounting, and that the SRC would not meet any advance expenditure in Britain other than that appropriate to the standing advance held in accordance with the financial rules.

The long process of negotiation came to a successful conclusion when Jones was able to inform Hosie that the proclamation of the Act had appeared in the Commonwealth Gazette and that formal notification from the Australian Minister to the Secretary of State for Education and Science, Margaret Thatcher, was on its way. The Anglo-Australian Telescope Agreement, the Act and the supporting regulations came into effect on 22 February 1971. The seventh and last meeting of the JPC took place at that time, and its principal business was to transfer formally to the newly constituted Board all assets, property and contractual rights and obligations held by the SRC and DES on its behalf.

The text of the Agreement is reproduced fully in Appendix 1; here we give some of the salient points which have governed the operations of the Board since its formation. The two Governments are the contracting parties to the Agreement and operate through designated agencies, which in late 1989 were the Science and Engineering Research Council in Britain and the Department of Employment, Education and Training in Australia. Since 1970 restructuring in the bureaucracies, especially in Australia, has seen the responsibility for the Board move from different departments of Government. Each contracting party appoints three members to the Board, one from each side being nominated to represent the corresponding Government agency on matters of financial and administrative concern. Articles 2 to 5 of the Agreement express in precise terms the conditions outlined by the Australian Government in April 1967 for the construction and use of the telescope. In particular, Article 4 refers to the opportunities for arrangements with the Australian National University and was the often quoted Article in the period of disagreement between the Board and the University, which is described in detail in Chapter 8.

Articles 6 to 10 define the status of the Board and the conditions under which it operates. The Board derives its legal standing and powers under Australian law from the Anglo-Australian Telescope Agreement Act, 1970. It has ownership of the telescope, and somewhat like company directors who are accountable to their shareholders, Board members are answerable to the Governments which appoint and can dismiss them. Articles 11 to 16 cover the general financial framework within which the Board must operate and be accountable. The Agreement foresees that, except in specially agreed cases, all costs will be shared equally between both countries. Finally, Articles 17 to 24 deal with conditions of employment of staff, referring to items such as the retention of pension rights and the entry into Australia of persons and goods required for the operation of the telescope.

With the Agreement and Act in place, there was now a permanent framework within which development of the project could proceed without frequent recourse to *ad hoc* arrangements. However, negotiations continued about the interpretation of the Agreement and, in particular, of the financial rules. The main problems centred on the organisation for the management and operation of the telescope, on the relations between the Board and the Australian National University, and on the maintenance of the bi-national character of the whole operation. These are described in later chapters.

Notes to Chapter 4

1 Hoyle F. (1982). *The Anglo-Australian Telescope* University College Cardiff Press
2 Letter E.G. Bowen to W.L. Francis (SRC), 4 July 1967: 'When possible, it is sound practice and in the end most economical to place a prime contract for the construction of a large instrument of this kind. We did this at Parkes with considerable advantage to CSIRO. However, it is already clear that this will not be possible in the case of the 150-inch telescope. The blank will almost certainly come from the USA, the grinding may be done in the UK or Europe, the mount will probably come from the USA, the building and dome could well be designed in Australia, and so on. In other words, the agency which will ultimately be given authority by JPC to build the telescope must act as its own prime contractor.'

5 Site, dome and building

5.1 Siding Spring Observatory

One of the most important decisions to be taken with respect to an astronomical telescope is where to put it. As telescopes have become more efficient over the years, so has pressure increased to locate them at sites where they can be used to the best advantage. Clearly a good site is one where a minimum of time will be lost through cloud, and where there will be a minimum of interference from city and other lights. Other less obvious factors enter, isolation and staff living conditions among them, but by far the most important is that known to astronomers as 'seeing'.

Seeing plays an important role in this book. It is a meteorological phenomenon, akin to the the shimmer observed on a hot day, and its effect is to blow up a star image – a point – into a blur or disc, the seeing disc. It is caused by turbulence, the irregular mixing of masses of air at different temperatures, which may take place at any level between the upper atmosphere and the dome itself. The differing refractive powers of these air masses deviate the light rays in a random manner and hence degrade the images they form in the telescope. Seeing can have a most deleterious effect on the efficiency of a telescope, and bad seeing is a great bugbear of astronomers. Influenced by many factors, it can vary widely from site to site, and especially in recent years considerable efforts have been made to locate sites where the seeing is the best possible.

Siding Spring Observatory owes its existence to Bart Bok. Very soon after he arrived in Australia in 1957, to succeed Woolley as Director at Mount Stromlo, he saw how seriously the future of MSO was threatened by the rapid growth of Canberra. The immediate effect of this, or of the growth of any large city, is that street and other lighting increase the brightness of the night sky so much that many of its best-known features, such as the Milky Way, can no longer be seen; more importantly, astronomical observing efficiency is seriously impaired. Observatories in many cities have been affected by this so-called light pollution, and have been forced to find new locations. A well-known example was the move of the Royal Greenwich Observatory from Greenwich in London to Herstmonceux in Sussex.

Within a short time, and with characteristic energy, Bok set about finding a site where a field station for future MSO telescopes could be established without too much delay, and which might also serve for a possible large telescope; he had this in mind from the beginning. The criteria, on which we elaborate later, were good seeing, dark skies and reasonable freedom from cloud. The final choice, that of Siding Spring Mountain, was made in May 1962, and by 1967 Siding Spring Observatory was well

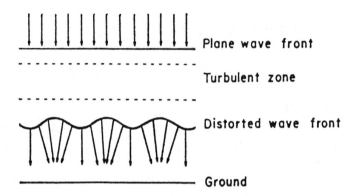

Figure 5.1 Distortion of a wavefront after passing through a zone of atmospheric turbulence

established. It was the obvious place to locate the Anglo-Australian Telescope, and the decision to do so was accepted without demur.

Bok's site survey was an ambitious undertaking. Initially it was confined to New South Wales – logistics dictated that any ANU field station be within a day's drive of Canberra – but as the prospect of a large telescope became more real, it was extended to sample the southern half of the whole continent. The further south, the more favourably could the Magellanic Clouds be observed, and southern Australia was not affected by the 'wet', the heavy monsoonal rains which blanket the north in summer. Half of Australia is a considerable area, equivalent to half of mainland USA or much of Western Europe. Much travelling was involved, and staff members who took part in the site programme found themselves becoming familiar with parts of the great Australian outback they hardly knew existed.

The immediate conduct of the survey was in the hands of A.R. Hogg, of the MSO staff.[1] Potential sites were selected from a combination of meteorological and topographical data, and survey parties were sent out to check at first hand ground configuration, access, distance from the nearest town, and the availability of power, water and telephones. The next step was to gather local meteorological data, with emphasis on cloud cover, and if the site still looked promising an observer would pay it periodic visits of a few weeks at a time, during which he would make systematic measurements of seeing, atmospheric transparency and wind velocity. It was a time of great activity in southern hemisphere site testing with surveys being conducted in South Africa and South America as well as Australia. Much interest was taken in the MSO programme, and three American institutions, the Yale–Columbia Observatory, the University of California, and Mount Wilson Observatory (CARSO), sent their own expeditions to Australia, all co-operating closely with the Mount Stromlo group.

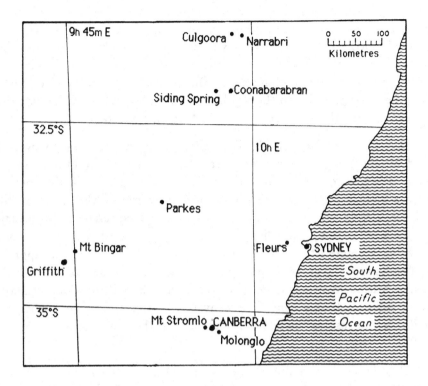

Figure 5.2   Map of New South Wales showing the locations of the major observatories

Siding Spring was first suggested as a possible astronomical site by Harley Wood, the New South Wales Government Astronomer, and an astronomical party first visited the district in August 1957. The party comprised Harley Wood; Richard Twiss, who needed a site for the University of Sydney's intensity interferometer, and found one in Narrabri; Isadore Epstein of the Yale–Columbia Observatory, looking for a site for a proposed 20-inch astrograph refractor (which went to Argentina, then to Chile); and Ben Gascoigne of Mount Stromlo Observatory. The party went first to Narrabri, primarily to inspect Mount Kaputar (1508 metres), then to Coonabarabran. At that time it was thought that Siding Spring was only 800 metres high,[2] and access to it being difficult, the party decided to bypass it. They were entertained by the Shire Council to a memorable dinner which included such delicacies as Roast Turkey Virginis, Sirloin Tauri and Coffee Milky Way, then pushed on south and west.

Some time later Theodore Dunham, of the MSO staff, discovered that the true height of Siding Spring was more like 1200 metres. He and Gascoigne paid it a second visit, reaching it by way of a little country road crossed by a dozen gates, followed by a climb through the scrub; it was a far cry from the rather grandly named Renshaw Parkway the road has since become. Gascoigne well remembers standing on the knoll where the satellite camera was later located, surveying that long, indented, northward-tilted scarp, trying to assess how efficient the air drainage would be, and to imagine, without much success, how a few telescope domes would look against that unforgettable landscape. Dunham and he reported favourably to Bok, Siding Spring was added to the regular observing list, and early in 1960 the site testers moved in with their telescopes, test instruments and tents.

By 1961 the choice of New South Wales sites had been narrowed to two, Siding Spring and Mount Bingar. Bingar was located just north of the Murray Irrigation Area near a little town called Yenda, and had been occupied on a full-time basis since 1959, when a 26-inch telescope and a house had been erected there. It had been a successful site, clearly superior to Stromlo, and had yielded some good astronomical results, especially for Bok, who when the time came was most reluctant to leave it. But the evidence pointed to better seeing on Siding Spring, which was appreciably higher and had a more favourable configuration, with high steep cliffs on its north and west faces. A brisk controversy developed, in which Bok and Hogg found themselves on opposing sides. It took a conference convened by the ANU Vice-Chancellor, Sir Leonard Huxley, to resolve the issue. He announced the choice of Siding Spring on 12 May 1962.

Much had to be done, and Bok threw himself wholeheartedly into developing his new observatory, which grew apace. Its first telescope, a 40-inch of advanced design, had been ordered from the American firm of Boller and Chivens in 1961 and was due to arrive late in 1963. It was, incidentally, the first really up-to-date telescope MSO had had (the 74-inch was very much a pre-war design). Before it could be installed, a road to the top had to be built, power and water supplies laid on, and a building and dome erected for the telescope. Everything was ready in time, and the first astronomical plate with the new telescope was taken by Gascoigne on 12 February 1964. Then came two smaller telescopes (16- and 24-inch), followed by a motel-type lodge for visiting astronomers and a number of houses for resident staff. The advent of the AAT, then of the UK Schmidt, and the ANU 2.3-metre in its rectangular building, led to sweeping changes. Siding Spring had to move quickly to accommodate these great instruments and their accompanying staffs, and within a few years it had grown into a world-class Observatory. Not even Bart Bok, most sanguine of directors, could have imagined such a transformation in so short a space of time.

Siding Spring Mountain by road is 30 km west of Coonabarabran, 500 km north-west of Sydney, and 660 km due north of Mount Stromlo. Not a conspicuous object itself – it is more a ridge than a mountain – it forms part of the eastern boundary of the Warrumbungle National Park, a place of spectacular natural beauty. One of the Park's characteristics is a series of remarkable volcanic plugs over a hundred metres high, the rocky cores of long-extinct volcanoes, exposed by millions of years of

Figure 5.3   The Warrumbungles Ranges seen from the Observatory

weathering and erosion. The Park and the Observatory together form a major tourist attraction which draws some 80,000 visitors a year to Coonabarabran, and has quite changed the nature of the little town. Its commercial interests now centre around motels and tourist buses rather than wheat and cattle, and visiting astronomers are no longer objects of intense interest and some suspicion.

5.2 The earth's atmosphere

The importance of seeing and associated atmospheric phenomena prompts a brief discussion of the subject. After all, it is to escape the effects of the earth's atmosphere that man goes to the considerable trouble and expense of putting his telescopes into space. We have noted that seeing is highly variable, from place to place and from time to time. Extensive site-testing programmes, guided by more or less empirical principles, have been conducted on every continent. The most successful sites have been found on high land, and are often isolated peaks, rising out of desert country or out of the sea, and high enough to be above local fog or mist, or dust storms. Proximity to lofty mountain ranges which can force passing air masses to shed their water vapour is also an advantage. The best-known sites include those occupied by CTIO and ESO on the southern edge of the Atacama Desert in Chile, by the telescopes on the peak of Mauna Kea, on the island of Hawaii, and by the telescopes on La Palma, in the Canary Islands. Their skies are dark, with 80 to 85 per cent of their nights usable, as compared with 60 per cent at Siding Spring, and their average seeing is probably better than that at Siding Spring, though excellent seeing has

Figure 5.4   Siding Spring Observatory in 1988 *(Photo: ROE)*

been experienced at the latter. Other famous sites, like Mount Hamilton and Mount Wilson in California, are now seriously affected by light pollution.

Because Australia has been denied the advantage of a high mountain range, it is possible that it contains no sites of this highest class.[3] Nevertheless the acid test of scientific productivity demonstrates that Siding Spring has clearly been a successful site, and indicates that factors other than climatological ones may be more important than have been realised. For instance, Australia is more advanced technologically and more stable politically than other countries in the southern hemisphere. We return to this subject in Chapter 14.

The earth's atmosphere has other effects. Because it is slightly refractive it acts as a weak negative lens, making celestial objects appear closer to the zenith than they really are. Thus 45° from the zenith refraction moves a star some 60 arc seconds higher in the sky, and a telescope pointing at it must be adjusted accordingly. Refraction at the horizon is about half a degree, so that the rising (and setting) sun and moon are displaced upwards about a whole diameter from their true positions. Refraction is colour dependent, greater in the blue-violet than in the red, and 60° off the zenith a star image is drawn out into a little spectrum one arc second in length. For stars less than 20° above the horizon the effects of refraction are serious enough to compromise significantly the efficiency of a telescope.

Again, even the clearest of skies absorbs light, or more accurately scatters it. Before

Figure 5.5   Two plates of Centaurus A showing good and bad seeing taken with the AAT

the light from a star in the zenith reaches the earth's surface, it will have lost about a fifth of its red light and two-fifths of its blue light in this way, not to mention much of its infrared and the whole of its ultraviolet below 0.3 microns, the latter being absorbed by the ozone layer. Light absorption increases with zenith distance because the mass of air the beam has to traverse also increases with zenith distance. For the same reason the seeing at lower altitudes is inferior to that at higher.

As a clear sky absorbs light, so does even the darkest night sky emit light. It is made up of airglow (akin to the aurora) from the upper atmosphere, zodiacal light scattered from dust in the solar system, and the accumulated emission from faint stars and galaxies. This light is important, because faint though it is, it is the brightness of the night sky, together with the seeing, which jointly determine the faint limit below which a telescope cannot work effectively. Figure 5.5 is a juxtaposition of two exposures on the same galaxy, Centaurus A, one taken in good, the other in bad seeing. Obviously individual details stand out more clearly and can be observed more efficiently when the seeing is good.

With a little algebra the argument can be made more precise. The problem is classical, the detection of a signal in the presence of background noise. Consider how we might measure the brightness of a star with a ·detector which could be a photographic plate, though a photoelectric cell (as in the exposure meter of a camera) or one of the more recent electronic detectors would be better. The telescope has diameter D, and light reaches the detector after being isolated by a small focal plane diaphragm, the diameter of which, β, is chosen to match the seeing. One first centres the star in the diaphragm, and measures the light from the star S together with that from the background B for a time t, the total signal being $(S + B)t$. One then moves to an adjacent area of sky and measures Bt. We must bear in mind that light is not a continuum but arrives in little pieces called photons (they resemble raindrops in some ways). Elementary statistics then give for the signal and its uncertainty

$$(S + B)t - Bt = St \pm [(S + 2B)t]^{\frac{1}{2}}$$

S and B are the numbers of photons received per second. S will depend on D^2, and we write $S = sD^2$. B will depend on β^2 and D^2, and we write $B = b\beta^2 D^2$. Then the signal/noise ratio is

$$St^{\frac{1}{2}}/(S + 2B)^{\frac{1}{2}} = sDt^{\frac{1}{2}}/(s + 2b\beta^2)^{\frac{1}{2}}$$

There are two limiting cases, $B << S$ and $S << B$: in the first the star dominates, in the second the sky. Considering only the second, suppose we are aiming at a signal/noise ratio of R; R might be 10/1 for a difficult object, corresponding to an accuracy of 1/10 or 10 per cent. Then

$$R^2 = s^2 D^2 t / 2b\beta^2$$

or

$$t = 2b\beta^2 R^2 / s^2 D^2$$

This makes clear the importance of dark skies, efficient sky subtraction and good seeing. In the present case, although Canberra is hardly noted for its bright lights, and Mount Stromlo is 15 km from the city centre, the night sky at the Observatory is perhaps ten times brighter than that at Siding Spring. All else being equal, to obtain an equivalent result at MSO one must observe ten times longer than at Siding Spring. The importance of seeing will be equally obvious: if the seeing deteriorates by a factor two, β is doubled and the observing time t must be increased four-fold. Note also that good seeing means good definition, for many problems a crucial advantage. And while in a given time, and other things being equal, a 160-inch cannot go fainter than a 40-inch telescope by more than a factor of four, if both are measuring the same star it will take the 40-inch 16 times as long to reach the same accuracy. Alternatively, the 160-inch could measure 16 similar stars while the 40-inch was measuring one. We can also see why it is so important for astronomers observing faint objects to work in the dark of the moon. At the blue end of the spectrum, the sky is 50 to 100 times brighter at full moon than it is at new moon (though the difference at the red end is much smaller).[4]

This is the simplest case, which applies only to faint stars and in which the only noise source considered is the photon noise of the signals. With a spectrograph for instance, one would consider also the nature of the source, point or extended, and the size of the collimator or dispersing element. The conclusions would then differ in detail but follow the same general lines.

5.3 The site

The choice of site brought a number of important questions into focus, and when the Joint Policy Committee (JPC) met for the first time in August 1967, it had before it a paper from the Technical Committee listing five matters on which immediate action was needed. To the surprise of some members of the JPC, the list was headed not by the blank for the primary mirror, but by the combination of site works, dome and building.[5] There were in fact well-known cases of telescopes, the University of Hawaii's 84-inch, the 92-inch at Steward Observatory, Arizona, and 74-inch at Mount Stromlo among them, which had been completed well ahead of their buildings. The JPC had already offered long-term contracts to John Pope and Herman Wehner. It now gave Wehner the main responsibility for the site, buildings and dome, while Pope took immediate charge of the telescope tube, mounting and drive.

Several problems required urgent answers. At which site on Siding Spring Mountain was the AAT to be located? To what extent did the mountain-top facilities need to be upgraded? What would be the main dimensions and requirements of the building and dome? The last had to be known before a design study could proceed. The diameter of the building is fixed by the diameter of the the dome, which in its turn is decided by the focal length of the primary mirror. The height of the building depends on how far the telescope is to be placed above the ground, and this is decided by astronomical site tests. Measurements, principally of microthermals, had shown that most sites had a layer of poor seeing near the ground, to avoid which the telescope needed to be raised anything from 12 to 30 metres or more above it. Opinions differed as to exactly how far, and local topography was an additional complication, but in the late 1960s the general view favoured heights of upwards of 30 metres.[6] This figure could not be decided until comparative tests had been made at the proposed sites, and the final location had to be selected before construction of the building and dome could begin. The question was a vexed one because such a large amount of money hung on it.

The final choice lay between site 3A, at the western end of the ridge, at its highest (1164 metres) and most exposed point, and site 7A, which was at the eastern end,

near the observers' Lodge and at an altitude of 1135 metres. The choice between
the two had to be decided by measurements of astronomical seeing, wind, dust and
humidity, together with a geological examination of the foundations. All testing was
carried out under the direction of Redman.

Not much credence was placed on conventional seeing tests conducted at ground
level, partly because of the influence of the ground layer. The principal cause of bad
seeing is thermal fluctuations, and as these could be measured more easily than seeing
itself, especially at heights above ground level, they were given most of the weight.
The air temperatures were measured with high-speed resistance thermometers with
sensors made of very fine wire, like electric light globes without envelopes. They were
mounted at intervals up masts some 30 metres high, and were of course exposed to
the air, and to occasional swarms of passing insects. There was trouble with strong
winds, which more than once blew complete instruments away, and with lightning,
which on one occasion destroyed a whole mast. Accurate wind results were finally
obtained; they were lower for site 7A, but microthermal and humidity readings showed
no significant difference between the two sites. The microthermal fluctuations at 33
metres averaged about 80 per cent of those at 18 metres, and Redman's preliminary
verdict was *'the figures so far suggest that . . . a centre of motion 70 feet above
ground . . . would be high enough at any of the three sites considered to avoid
significant ground effects'.*[7]

As planning for the AAT developed, other factors came to the fore. Wherever it
was, the site had to accommodate a large air-conditioning plant, a stand-by power
generator, and other ancillary services, as well as a tourist building and a tourist
car-park. The difficult and limited access to site 3A and the restricted space around
it were becoming real drawbacks. Core samples drilled at the two sites showed that
3A was the better, although adequate foundations could be found at 7A some 1.5
metres below ground level. But 3A had no other advantages, and it was no surprise
when the decision went to site 7A.

While local site testing was proceeding, the Project Office had to decide on the
main dimensions of the telescope so that building and dome design could begin in
earnest. The decision on the optical design was the first stage of this process. A
considerable upgrading of the essential services on the mountain was also necessary,
and all this had to be completed before construction could begin.

A first concern was to improve the road so that heavy and sometimes unwieldy
loads could reach the Observatory without too much trouble. The little road Dunham
and Gascoigne had driven along some ten years before was now the access road to
the National Park, which is to say that the 13 gates had been eliminated and the
first 14 kilometres sealed. The JPC arranged for a bypass to be built around the
Blackburn Hill sector, notorious for an acute elbow bend around which passengers

often preferred to walk rather than ride. Then, defeated by the procrastinations of the various public bodies concerned, the JPC found itself having to fund the sealing of the nine kilometres that remained between the bypass and the Observatory turn-off. Including the bypass, the upgrading of this road cost about $230,000. As the road also gave direct access to the National Park from Coonabarabran, these improvements were a considerable windfall for the Coonabarabran Shire Council.

The Observatory site was of course the property of the University, and the improvements there, from which both parties stood to gain, were carried out by the University under an arrangement with the JPC. The total cost was about $1.4 million, of which the JPC met 60 per cent. The road up the mountain was re-aligned and widened, while the mountain-top roads were resurfaced and sealed. A tourist centre was erected, and a large car-park created adjacent to it. Five new houses were built, and the observers' Lodge more than doubled in size. A large utilities building was erected close to the telescope, to house the AAT's air conditioning plant, a diesel-driven 400 kVA standby generator, a mechanical workshop, and other equipment, mostly electrical.

To increase the water supply meant laying a pipeline from the town water supply at Timor Dam, about half way between Coonabarabran and the Observatory. It took four pumping stations to raise the water to the top of the mountain. In its first years the Observatory telephone was connected to the outside world by a temperamental radio link between the mountain top and the Coonabarabran Post Office. Some observers enjoyed the comparative isolation this conferred, but it vanished abruptly when the radio link was replaced by a 30-pair underground cable. As was soon demonstrated, it was now possible to make a call from the prime focus cage of the AAT direct to the Royal Observatory, Edinburgh. At the same time the capacity of the 22 kV electricity supply to the mountain was greatly increased.

5.4 The dome and building

Astronomers have long found it difficult to understand why the dome and building of a large telescope should often cost as much as the telescope itself. Like the Kitt Peak and ESO telescopes, the AAT was no exception to this rule, but at least there was something to show for the money. With a height of 50 metres and a diameter of 37 metres, the AAT building is by far the largest in northern New South Wales. It is a handsome building, and even in that region of incredible skylines its gleaming white dome makes it a commanding object for many kilometres around.

Preliminary work on the dome and building began in 1967. Expert assistance was necessary, and in March 1968 the services of a Sydney firm of consulting engineers, Macdonald Wagner and Priddle, were enlisted for the purpose. The firm was well-known in Australian technical circles for its work on the Parkes radio telescope, and it was employed also by the ANU to superintend the mountain-top developments.

A general layout of the building was approved in March 1969, clearing the way for tenders to be called. However, there were financial problems: estimates continued to rise, and the first tenders came in disconcertingly high. Money-saving modifications had to be made, and the scheme for the large-scale circulation of air through the dome at night was postponed. Fortunately the ducting and the building modifications for locating the big fans were retained. Some years later, in response to persistent dome seeing problems, the air circulation scheme was restored much as originally planned. The main building tender was awarded to Leighton Contractors Limited of Sydney in October 1970. The dome was sub-contracted to Evans Deakin Limited of Brisbane, and the total cost had risen from an initial estimate of \$3.2 million to \$4.65 million.[8] The dome and building were effectively complete by the end of 1972.

The AAT and Kitt Peak telescopes being so similar, certainly in their external features, it was natural for the AAT to follow the design of the Kitt Peak dome and building. The main features of the AAT building are shown in Figure 5.6, a simplified sectional drawing. Like the building for the Kitt Peak Telescope, the AAT building is erected around a circular concrete pier, which carries a hammerhead or flat cantilevered top on which are mounted the telescope and the coudé instruments. The hammerhead also is of concrete, 1.2 metres thick, and the pier is a cylinder of external diameter 10.25 metres with a wall 30 cm thick. The hammerhead and pier are built on an independent foundation, and are separate from the building at all levels. This isolates them from vibrations of the building, in which respect the design has been most successful. The pier also houses the primary mirror elevator.

The Kitt Peak building is made of steel, but concrete was preferred in Australia; concrete is cheaper there and has other advantages, for instance, in its thermal properties. The declination axis of the AAT is 29.4 metres above the ground. Ten years later it might well have been made lower, as Bowen would have preferred,[9] though to go much below 24 metres would have entailed considerable design changes. It was also suggested at one stage that a separate building be provided for the library, astronomers' offices, and measuring equipment, and this course was in fact followed when the William Herschel Telescope was erected on La Palma. Apart from the telescope itself, the WHT building is almost empty.

The 26-metre high AAT building contains seven principal floors. The ground floor was planned so that trucks could be parked directly underneath a hatchway 5.5 metres square, up which the big dome crane could lift loads to any floor. The largest load was the aluminising tank; the heaviest, at 46 tonnes, was the centre section of the declination axis; the most valuable was the primary mirror. The next floor contains the library, astronomers' offices, the electronics laboratory, and the main photographic section, all located as far as possible below the telescope because of the amount of heat they produce. The following floor houses the aluminising plant with its massive vacuum tank and pumps, and the hydraulics for the oilpad bearings,

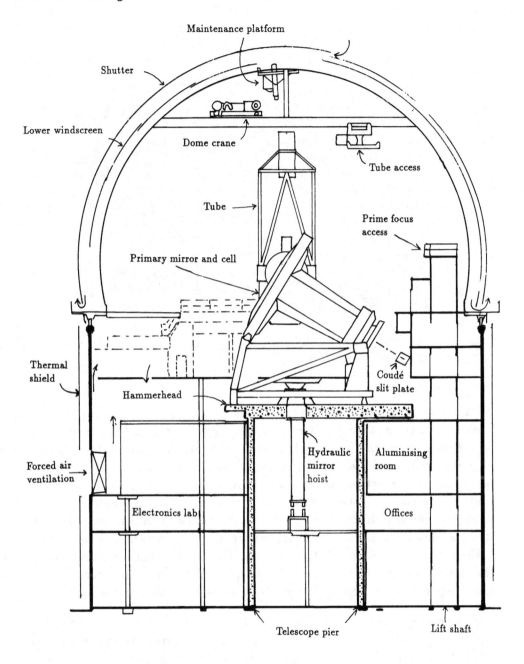

Figure 5.6 A simplified sectional drawing of the building and dome

Figure 5.7   The domes and buildings housing the AAT and the Schmidt Telescope (foreground) on Siding Spring Mountain

including the oil reservoir, pumps, filters and coolers. On the same level, but installed some time later, are the large fans for ventilating the dome, and the newer computers. Above this is an insulating floor planned as a thermal barrier against any upward flow of the heat from the lower levels. Then the main floor, which includes the telescope base, the coudé spectrograph and a large instrument testing room. This floor also contains the visitors' gallery, which has its own access. Finally there is a small floor mostly taken up with the screened electrical junction room, and immediately above it is the console room from which the telescope and its instruments are controlled; it is the nerve-centre of the whole operation. The console floor is on the same level as the catwalk outside the dome (see Fig. 5.8), to which it has direct access. The prime focus cage is entered from a platform built above the console room roof.

The primary function of the dome is to keep out the weather and to serve as a windshield, wind-induced vibration being fatal to observing. Wind can also cause the whole building to shake, as can dome rotation, hence the care that neither of these motions be transmitted to the telescope. The dome is a considerable structure. It weighs 570 tonnes, as much as a small train, and like a train it is mounted on bogies – 32 two-wheel sprung bogies, which are attached to a horizontal ring girder and run on a circular steel rail, ground flat. The rotary motion has to be smooth, accurate and easily controllable, as must the motions of the shutter and the windscreen. The dome must be strong enough to support a 46-tonne capacity crane, essential for the

Figure 5.8   The AAT with native blackboy plants in the foreground

initial assembly of the telescope and for handling the primary mirror when it is being aluminised. Two massive parallel arches mounted on the ring girder carry this crane. They also support the dome shell and shutters and define the 5.3-metre viewing slit through which the telescope sees the sky.

The assembly of the dome presented some problems, and it could not be carried out until virtually the whole of the space from the main floor to the underside of the dome 26 metres above it had been filled with a forest of scaffolding (see Fig. 5.10). Its main purpose was to hold the arch segments in place while they were aligned and welded. The arch supports were then withdrawn but the scaffolding was retained, to provide access while the long intermediate panels for covering the dome were lifted into place by crane, and while the dome lighting and insulation were installed. There were heart-stopping moments when panels were sometimes caught and spun around by sudden gusts of wind, but everything finally emerged unscathed.

There was one bad accident. It happened when a temporary platform near the top of the building gave way, and caused an unfortunate workman to fall to his death. This so shocked his workmates that within a short time all had left and found jobs elsewhere.

High winds being a problem on Siding Spring, it was decided that the telescope should look through a square aperture rather than through a slit, the aperture being defined by an up-and-over shutter and a windscreen. The aperture is 5.3 metres

Figure 5.9   The under side of the dome showing ring girders and bogies which allow the dome to rotate

Figure 5.10   Erection of the dome. Note the forest of scaffolding and the partially erected main arches behind the crane *(Photo: Don Collins)*

Figure 5.11   Guiding into position one of the last panels of the dome *(Photo: Don Collins)*

square, and as it clears the incoming circular beam by only about 50 cm on each side, the dome and windscreen must be kept well aligned with the telescope. However, no wind vibration has been noticed even on nights when other telescopes on the mountain have been obliged to close. There were mechanical and electrical difficulties in setting all this up, but the real problems proved to be those associated with the temperature of the dome air. Dome air is heated by solar radiation, by the computers and the racks of electronics located about the telescope (a major and growing source), by heat stored in the telescope, especially if the mirror warms up during the day, and by upwards diffusion from heat sources on lower floors. These effects set up temperature differentials which, as in the atmosphere itself, create turbulence and can affect the seeing markedly. The air outside the dome complicates matters because it heats up during the day, then cools rapidly around sunset, levelling out later to a greater or lesser extent.

In designing the dome, Wehner set out to create a regime in which, ideally, the dome temperature would be allowed to rise slowly during the day in such a way that when the shutters were opened at nightfall the air temperature inside would be equal to that outside. The aim was then to maintain this equality throughout the night. The problem of solar heating was met in the usual way by cladding the dome with an inner and an outer skin, between which fans installed around the base of the dome could circulate air from the outside, so that it flowed either from top to bottom or vice versa (see Fig. 5.6). The inner skin was insulated, and the system worked well, with daytime thermal gradients inside the dome kept small and well under control. The night temperature inside the dome was to be kept in step with that outside by a set of powerful fans with heaters and coolers, all installed on the second floor. Working through an associated system of ducts beneath the main floor it would replace the dome air about once every five minutes. A further problem concerned the mirror itself. The mirror has a long thermal response time and is a considerable source of heat, which it tends to dissipate into the layer of air immediately above it. A system of fans and jets was installed around the mirror cell to disperse this layer, and generally to keep the temperature gradient between the mirror and the outside world as smooth as possible. This was the first time active air control had been proposed on such a scale, and at the time the system was described as *'far beyond anything installed in previous observatory buildings'.*[10] It should be mentioned that Redman, always interested in such matters, made a substantial contribution to the scheme, as also did the CSIRO Division of Mechanical Engineering, which assisted with some special studies of heat flow and other problems.

Unfortunately, for the financial reasons mentioned above, the part of the installation concerned with the forced ventilation of the dome had to be postponed. It was several years before this main feature could be reinstated, and then not in its entirety. The relevant events are described in the next section.

Despite these problems, now mostly in the past, the building has proved efficient, handsome and a pleasant place in which to work. In 1974 it was awarded the Engineering Excellence Award of the Association of Consulting Engineers Australia.

5.5 Dome thermal control: epilogue[11]

Suspicions that the seeing on the AAT was deteriorating had been voiced as early as 1976. They were heightened by comparisons between image sizes estimated by AAT and Schmidt observers, according to which images observed with the 25-cm guide telescope on the Schmidt tended to be smaller than those seen with the AAT. It was also noted that the air temperatures inside the AAT dome often differed appreciably from those outside, with the inside usually warmer, at times by several degrees C. Moreover there was a distinct correlation between temperature differential and image size, with image diameters increasing by roughly 0.5 arc seconds for each degree of temperature difference.

In view of the relatively small dome aperture, the poor dome ventilation, and the fact that the forced ventilation originally planned had been abandoned, the existence of an air temperature differential was not surprising. A first step was to explore the microthermal properties of the air inside and at the boundary of the dome. This was done with fast-response, fine-wire resistance thermometers of the type used during Redman's site tests. They showed that, while the fluctuations inside the dome were quite small, those near the aperture could occur with amplitudes of 3 or 4°C,

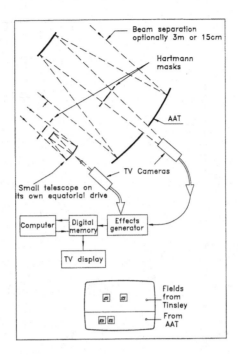

Figure 5.12   Gillingham equipment for comparing seeing from inside the dome with that in the open air

comparable with the temperature difference between the inside and the outside air, and consistent with a picture of bubbles of hot air breaking away at the interface. Thermal effects of this magnitude are enormous compared with what happens in the open air, and a rough calculation showed they could well be responsible for the suspected image degradation.

To confirm these ideas Gillingham devised a direct, quantitative, real-time method of discriminating between dome seeing and external seeing, the first time it had been done. It was a variant of the Hartmann test for telescope objectives, and the principle is indicated in Figure 5.12. A screen is placed over the entrance pupil of a small test telescope (20 cm aperture) which isolates two holes of diameter 5 cm, with centres 15 cm apart. The telescope is pointed at a bright star, and the star is detected with a television camera placed far enough out of focus for the beams from the two holes to form separate images. About once every two seconds the positions and separation of the image centroids are measured electronically. The separations are averaged over a series of frames, and the amount by which they vary – the standard deviation of the series – is taken as a measure of the seeing. This measure is not affected either by fixed aberrations of the telescope or by small irregularities in its drives, including vibration.

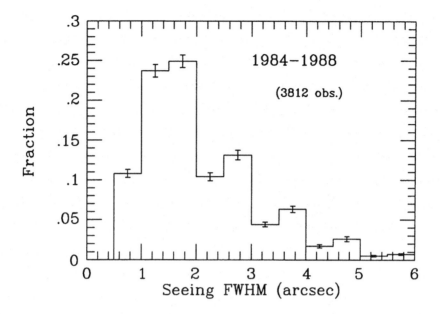

Figure 5.13   Statistics of five years' seeing in the AAT

An analogous screen defines analogous beams in the 3.9-metre telescope, and the observation is arranged so that the pairs of images in the two telescopes are detected and processed simultaneously. In a typical experiment the first run is made with the test telescope mounted on the catwalk outside the dome where it can measure the atmospheric seeing directly. The telescope is then moved inside and observes out of the dome aperture, measuring now the combination of atmospheric and dome seeing. The simultaneous observations made with the 3.9-metre serve as a control. In this way it is possible to demonstrate the deleterious effects of dome seeing quantitatively and unambiguously.

Gillingham concluded that one could reasonably aim to match the air temperatures inside and outside the dome to within 1°C, and that the most cost-effective way of doing this, certainly as a first step, was to increase the ventilation about four-fold, utilising much of Wehner's original forced ventilation scheme. Two large fans were therefore installed at the site originally planned for them on the second floor. With diameters of 1.9 metres and rated flows of 40 m³/sec, they could replace the dome air in about four minutes. Their effect was obvious, and during their first year of operation it was estimated that they increased the efficiency of the telescope by 35 per cent.

The technique has been adapted to study effects within the dome, like the ventilation of the primary mirror and the effects of temperature differences between the telescope steelwork and the dome air. A recent study of the seeing data made over the five years 1984–1988 yielded a median seeing of 1.8 arc seconds with virtually no year-to-year or season-to-season variation.[12] The study showed that the seeing tends to deteriorate during the night, especially when the dome air is colder than the mirror. Means for combating this effect should now be at hand.

Notes to Chapter 5

1 Arthur Hogg served as deputy to Woolley and Bok. A partial account of his site-testing work may be found in the biographical memoir, S.C.B. Gascoigne (1968). *Records of the Australian Academy of Science* 1, No 3, 65–67. No extended account of the site-testing programme is known to exist.

2 The height of Siding Spring Mountain was given in Bartholomew's 1953 *Advanced Atlas of Modern Geography* as 2817 feet or 859 metres.

3 Mount Woodruffe, 1440 metres high, 26.20°S 131.42°E, about 100 km south of Ayers Rock, looked as promising as any, but was too isolated and difficult of access for further consideration.

4 The numerical data in this section are from Allen C.W. (1973). *Astrophysical Quantities* 3rd ed. 134, §61.

5 First meeting of the JPC, 15–18 August 1967, minutes §12 'Several members of the Committee expressed surprise at these conclusions, and it was thought that in practice the mirror was on the critical path.'

6 See, for instance, *The Construction of Large Telescopes* (1966) IAU Symposium No 27, D. Crawford (ed.) Academic Press. B.H. Rule, 175, 'recent tests indicate for most sites . . . that it is desirable to have the dome at a much higher elevation above the ground, anywhere from 40 to 150 feet (12 to 50 metres)'. C. Fehrenbach, 165, 'most astronomers agree . . . at least 25, or even 50 metres for large telescopes'. D.L. Crawford, 181, 'Today [1965] our thinking is that we will be definitely 100 feet (30.5 metres) or more at the declination axis.'

7 From Technical Report No 3, presented to the second JPC meeting in March 1968: 'The detectors had a time constant of 0.015 seconds, and samples were taken once a second. The figures quoted are the r.m.s. of the differences between successive samples, taken over five minute runs. The mean fluctuation, averaged over height, was 0.038°C per second.'

8 Hoyle has some pithy comments on this point. *Anglo-Australian Telescope*, University College Cardiff Press, p.15

9 'My own view was . . . on a rugged site such as that at Siding Spring, already some 300 metres above the surroundings and with the telescope perched on the edge of a precipice over 100 metres high, height of the dome was not important. Compared with the cost of the telescope itself, the AAT dome was one of the most expensive built to that time'. Bowen's personal communication to the authors. (But domes were generally reckoned as about 30 per cent of the total budget – SCBG).

10 Third meeting of the JPC, March 1969, minutes §7.7

11 Gillingham P.R. (1983). *Proc. SPIE* **444**, 165 'Advanced Technology Optical Telescopes II'

12 *AAO Newsletter* No 48, (1989)

6 Optics and tube

6.1 The primary mirror blank

The central element of a large reflecting telescope is the primary mirror: decisions on it are fundamental and affect all aspects of the instrument. The mirror diameter fixes the size and light-grasp of the telescope (and often its name – the Russian 6-metre, the Palomar 200-inch); its weight determines the weight of the mirror cell, tube and mounting, and hence the engineering scale of the whole instrument; and its focal length fixes the size of the dome and building, and their cost. Not that the AAT group had much choice in the matter. The essential content of the decision that the AAT was to follow the Kitt Peak design was that the AAT was to have a mirror like the Kitt Peak mirror, and Kitt Peak had already settled for a solid monolithic blank with a maximum thickness near 64 cm and a finished weight approaching 20 tonnes. This was the current orthodoxy, and Kitt Peak, ESO and the Canadians all followed it, their thinking no doubt governed by the difficulties experienced with the Palomar 200-inch mirror. The saga of this mirror was well-known, from the costly early experiments with fused quartz to the adoption of pyrex, the difficulties of pouring and annealing the pyrex blank, and the protracted struggle to adequately support the the extremely flexible final mirror in the telescope.[1]

The development of low-expansion glassy materials had only recently made mirrors of the Kitt Peak type possible, and their massive proportions were an obvious answer to such flexure problems. The same massive proportions created problems of their own, in the very heavy mountings needed to carry them and the magnitude of the loads they imposed on the bearings, but they were conventional at the time and they worked. Ten years later mirror support systems had developed to a point where one would have confidently settled for a mirror with half the thickness; 20 years later one might have contemplated a ribbed mirror, possibly of pyrex, half as light again. Such possibilities were already being investigated in 1967, but the experiments had not been carried far and could not be considered seriously. Even if a suitable light-weight disc had been available, its adoption would have meant a complete reconsideration of the Kitt Peak design, which was out of the question.

The first priority of the newly formed Technical Committee of the AAT was to obtain a casting, or blank, for the primary mirror. Pieces of strain-free glass weighing 20 tonnes or more are not easily come by, especially when made of a low-expansion material like quartz, but with no less than three prospective purchasers – the AAT, AURA (for a telescope in Chile) and the French – and three manufacturers, it was for once a buyer's market. Of the manufacturers, the Corning Glass Company had for some time been synthesising silica by burning silicon tetrachloride with water vapour,

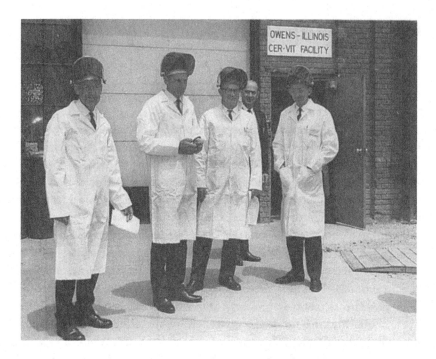

Figure 6.1   The AAT Technical Committee at the Owens–Illinois plant at Toledo, Ohio. From left Redman, Pope, Gascoigne and Wehner in white coats, with a member of Owens–Illinois behind. *(Photo: Owens–Illinois Inc.)*

and had already supplied large blanks to ESO and the Canadians. The General Electric Company had carried out much experimental work on quartz for the 200-inch telescope. Its method was to fuse crystals of natural quartz in a vacuum, and it had made the blank for the KPNO telescope. Owens–Illinois Inc. manufactured vast amounts of glass for the American domestic market. Its research and development division had developed a new product, a so-called glass-ceramic, by fusing silica with lithium, aluminium and other oxides, then subjecting the mixture to a heat treatment which left it part micro-crystalline and part amorphous. The resulting material was known commercially as Cervit, and its critical property, an almost zero expansion coefficient, could be demonstrated in spectacular fashion.

It had long been a dream of astronomers to have mirrors of low-expansion materials. If a mirror is heated or cooled unevenly, as by a rapid change in the weather, it will expand or contract unevenly and its surface will distort, perhaps by enough to make it unusable. The more the material changes with temperature, the more pronounced the effect; the mirror of the Mount Wilson 100-inch, made of relatively high-expansion commercial glass, could be put out of action for many hours by a rapid change in ambient temperature. With the 200-inch telescope the problem promised to be even more acute, and during the 1930s great efforts were made to avoid it by having the

Figure 6.2   The AAT Technical Committee wearing protective headgear while inspecting a 40-inch mirror blank which had just been poured at the Owens–Illinois plant. *(Photo: Owens–Illinois Inc.)*

mirror made of quartz. However, the high melting point of quartz created insuperable problems and the idea had to be abandoned. Instead, the mirror was made of a special low-expansion pyrex, in a honeycomb configuration of four-inch thick ribs designed to keep the thermal response time as short as possible. In this it was successful, though only at the cost of flexure problems discussed later. It was KPNO which encouraged industry to make a renewed attack on the problem of low-expansion glass, to the great benefit of astronomers everywhere.

All three manufacturers were clearly worth considering, and in July 1967 the Technical Committee visited their respective plants, all in north-eastern USA. The competition was keen, the hospitality lavish, and the tour had distinct alcoholic overtones. The styles of the companies differed markedly. Owens–Illinois had a large impressive research laboratory by which it set great store. On the other hand, Corning was staffed, as Redman described it, by *'practical engineers rather than physicists and had acquired great facility in the handling of fused quartz by long continued experiment rather than by laboratory research'.*[2] Owens–Illinois would pour a whole blank in one operation – a spectacular sight – whereas Corning and General Electric built up theirs piecewise by fusing together arrays of hexagonal 'boules' or ingots in a vacuum chamber. The Technical Committee discussed at length such problems as bubble content, internal strain, inhomogeneity and lack of dimensional stability, all

Figure 6.3   The Earl of Rosse and assistants, also wearing protective headgear, pouring the
speculum mirror for the Great Melbourne Telescope at Birr Castle in the heart of Ireland.
This happened almost exactly 100 years before the pouring at Owens–Illinois. *(Photo:
taken from a 19th century woodcut which appeared in* Strand Magazine*)*

of them important for the final mirror. Because Cervit, as opposed to silica, was new
and relatively unknown, it presented a particular problem: how stable would it be in
the long term? Redman had come prepared for this, armed by the physical chemists
of his home University of Cambridge with a truly formidable array of questions.[3]
But the Owens–Illinois physicists rose to the occasion in style and had an answer for
everything. The Technical Committee finally settled for Cervit and never regretted
it.

The decision had an interesting historical overtone. Almost exactly a hundred
years before, the city of Melbourne, flushed with the euphoria of the great Victorian
gold rushes, decided to purchase a telescope – a 48-inch telescope no less, the biggest
equatorial in the world. This, the Great Melbourne Telescope, was designed by a
committee in Britain which had had to decide, among other things, whether the
mirror was to be of speculum metal or of the new silver-on-glass type. Speculum,
the traditional material, was an alloy of tin and copper; it took a good polish but
tarnished rather rapidly, and when it did it had to be refigured, a difficult operation.
The silver-on-glass process had not long been invented, but once a glass mirror was
figured it could be recoated with little trouble. The committee too was made of
traditional material and chose speculum. Its decision was a near disaster. It was
a major obstacle to the efficient functioning of the telescope, and a hundred years

later could still remind at least the Australian part of the Technical Committee that history must not be allowed to repeat itself.

The three manufacturers had each submitted three quotes, depending on whether they sold one, two or three mirrors. Since the more they supplied the cheaper they would be, the JPC had to induce Kitt Peak and the French to follow its lead. This it did at the 1967 meeting of the IAU General Assembly, held in Prague. The JPC was the first organisation to purchase a large Cervit blank, and the second to complete a large mirror of low-expansion material.[4] The blank cost $US535,000. Since the JPC had expected the price might have been twice that, it was a good start.

6.2 The optical design

Optical design is not a glamorous subject and tends to be left to specialists. Nevertheless it is the stuff of which telescopes are made, and it was the next priority on the Technical Committee's list. The key points to be decided were:

1. the focal length, or focal ratio, of the primary mirror

2. the shape of the primary mirror – paraboloid or hyperboloid

3. the design of the corrector lenses for prime focus photography

4. the focal lengths and layouts of the various secondary foci

There is nothing subtle about these points, but the arguments are specialised, and will only be sketched here.

As already described, it is the focal length of the primary mirror which determines the size of the dome and hence the size of the building. As the dome and building can cost as much as the whole telescope large sums of money are involved. This was a central issue with the 200-inch, when the need to keep the dome cost within bounds led to the adoption of the then unprecedentedly low figure of f/3.3 for the focal ratio of its mirror, or 16.8 metres for its focal length.

However, there is a limit, set by the asphericity of the primary mirror, as to how far its focal length may be reduced. Such a mirror is almost always a paraboloid or a hyperboloid, and its asphericity is measured as the amount by which it departs from the sphere which fits it most closely. It depends inversely as the the cube of the focal ratio, and so increases rapidly as the focal length decreases; the more aspherical a mirror, the more difficult it is to figure.[5] Here we work at the limits of technology. Few organisations possess the equipment and ability necessary for such an exacting operation, and to ask too much of them may well jeopardise the quality of the final mirror, on which all else depends. Here again KPNO led the way, in that the

KPNO opticians were the first consistently to figure large mirrors faster than f/3.0. The telescopes for Kitt Peak and CTIO were designed around primary mirrors with diameters near 4 metres and focal lengths of 10.7 metres (f/2.6).[6] But, whereas these mirrors were figured in-house where experiment was encouraged, the AAT mirror had to be made commercially, and none of the prospective manufacturers approached at that time was confident of figuring a mirror faster than f/3 to the required standard. The focal length finally adopted for the AAT mirror was 12.7 metres, a focal ratio of f/3.3. Although this looks a modest change from the Kitt Peak specification, it had important consequences.

Over the years astronomical photography in the southern hemisphere had been sadly neglected. There were important southern objects for which not even basic photographs existed, and the AAT was seen as having a major role in redressing this imbalance. However, photography at the prime focus of modern fast telescopes is not possible without corrector lenses. While an unaided mirror can produce perfect stellar images on-axis, the off-axis images assume an arrow-head shape known as coma, which becomes worse with increasing distance and also with decrease in focal length. The result is that, with a telescope as fast as the AAT or the Palomar 200-inch, the field for good photography at the prime focus is barely the size of a postage stamp.

Fortunately this problem can be rectified by a special lens system, placed a short distance inside focus. Such systems are known as coma correctors, or simply as correctors, and the three made for the AAT are shown in Figure 6.4. The advantage of the simplest system [Fig. 6.4(a)], a figured glass plate, is that it has only two air-glass surfaces. It was built to cover a field of about seven arc minutes diameter, which is enough, for example, for the spectroscopy of objects with small angular diameter. It has not found much application on the AAT, though a similar system on the ESO telescope in Chile has been used with success.

The system shown on Figure 6.4(b), with two adjacent lenses of equal and opposite power, is of a type introduced by the American F.E. Ross around 1930. The third [Fig. 6.4(c)], with three separated lenses, was first described by the British designer C.G. Wynne in 1968. The Ross correctors were central to the whole optical design of the 200-inch, in that it was their success which enabled the focal length of the primary mirror to be made so short.[7] All the correctors shown in Figure 6.5 were designed for the AAT by Wynne. The Ross design will work at a focal ratio as small as f/3 only if the primary mirror is a hyperboloid. Such a mirror was also a prerequisite for Wynne's original triplets, though he later produced a similar design for paraboloidal mirrors. The AAT triplet covers a field of 24 cm diameter or 60 arc minutes. The doublet covers a field of 9 cm or 25 arc minutes, but can be used over a wider wavelength range than the triplet and has better definition. The triplet in particular is now widely used on big telescopes, where it has created new possibilities

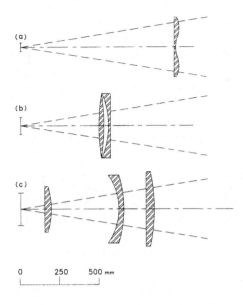

Figure 6.4   The prime focus correctors adopted for the AAT

for prime focus photography, and has stimulated much work in fields such as clusters of galaxies.

Focal ratios also had to be chosen for two Cassegrain foci (f/8 and f/15) and a coudé focus (f/30). The layouts are shown in Figure 6.5. The case for f/8 had gained the powerful advocacy of I.S. Bowen, famous for his association with the 200-inch. He showed that with the photographic emulsions available in 1965, and for reasons associated with sky brightness and optimum plate density, the deepest photographic exposures could be made at a focal ratio of about f/8. Such exposures would take from three to five hours. Another argument was provided by the Ritchey–Chrétien optical design. Proposed in 1922 and first tried in 1934, it did not really find favour until after 1960, when it was used by Kitt Peak for an 84-inch and by Mount Stromlo for a 40-inch. In this design the classical combination of a paraboloidal primary mirror with a hyperboloidal secondary is replaced by two hyperboloids, their exact shapes chosen to annul both spherical aberration and coma at the f/8 focus. A hyperboloid cannot be used at prime focus without a corrector, but the two mirrors together provide a wide f/8 field over much of which the definition is almost perfect.

Opinion remained divided for some time as to the relative merits of classical and Ritchey–Chrétien optics. Redman favoured the former, Gascoigne the latter. Classical optics were used for the Canada–France–Hawaii, Isaac Newton and William Herschel Telescopes, Ritchey–Chrétien for the ESO, KPNO and CTIO telescopes, and for the German Max-Planck-Institut telescope located in Spain. More recently

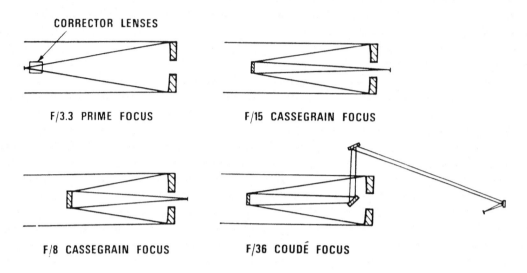

Figure 6.5   Optical configurations for the prime focus, Cassegrain and coudé foci

the success of multi-object spectrographs has tilted the scales in favour of wide-field systems like the Ritchey–Chrétien.

It was taken for granted that the AAT would have a coudé focus to provide the usual stable platform for equipment too heavy or otherwise unsuitable for mounting on the telescope itself; a high-dispersion spectrograph of the massive grating type was a common example. As the AAT engineers pointed out more than once to the AAT astronomers, the coudé focus cost a lot of money. It included one awkward mirror, the oval diagonal flat, 107 by 71 cm, which directed the light along the declination axis (Fig. 6.5). It was a difficult mirror to make, difficult to mount, and difficult to store out of the beam. The coudé position provides four separate locations where equipment may be left set up semi-permanently.

The f/15 focus was provided because certain instruments are easier to design for such a narrow beam, and because it would accommodate equipment already in use on other telescopes with f/15 foci. It had the advantages of being cheap and of enabling a quick interchange with the coudé focus. This was done by mounting the two secondary mirrors back-to-back so that they could be interchanged by a simple rotation.

This optical design, together with the recommendation to use Cervit for the mirrors, was accepted by the JPC at its meeting in August 1967. It was quite similar to the Kitt Peak design, possibly as a result of the common experience of both organisations with Ritchey–Chrétien telescopes. If it had a distinguishing feature it was a bias towards direct photography. How did the design fare in practice? First, within a couple of years the British Science Research Council had approved the building at Siding Spring of a southern counterpart of the Palomar Schmidt telescope, with the first task of mapping the southern sky as the Palomar Telescope had mapped the north. Second, the new Kodak emulsion, the celebrated IIIaJ, was not faster but slower than the old IIa and 103a, and could not be use at f/8. While it could be used, and with great effect, at the prime focus, a good deal of ground had been cut from beneath the AAT's photographic feet. Third, during the years that elapsed before the completion of the telescope, coudé spectroscopy went out of fashion, and demand for the already designed coudé spectrograph fell off to such an extent that it was not built. The coudé focus was used only occasionally by and microwave astronomers.

But all was not lost. There remained programmes which could be carried out in no other way than by direct photography, a situation which even the advent of sensitive electronic arrays like the charge-coupled device has not threatened.[8] David Malin, the scientific photographer at the Anglo-Australian Observatory, has pioneered new techniques to make a series of brilliant and valuable photographs which have been reproduced throughout the astronomical world, and have publicised the AAT as probably nothing else has. The f/8 focus has been taken over by a multi-object spectrograph which obtains spectra of 50 or more objects at one time, using optical fibres to channel their light onto the spectrograph slit. This instrument is revolutionising its branch of the subject, and for its purposes the good definition and 40 arc minute field at the f/8 focus are ideal. Fifteen years later a high-dispersion echelle spectrograph at last has been built for the coudé focus, though very different from that originally planned because it uses a detector with a very different format. Its success has been such that there are proposals to build an auxiliary telescope alongside the AAT solely to feed this spectrograph.

The mode in which a telescope will be used is highly unpredictable. It is at the mercy of ever-changing currents in astrophysical research programmes, and instruments and observing techniques are continually being supplanted by advances in technology which in turn make further programmes possible. As in the post-war decades some of the most critical attention was focussed in turn on Baade-type population studies, abundance determinations in brighter stars, and compact galaxies and radio sources, so observational emphasis passed from the prime to the coudé, and from the coudé to the Cassegrain focus. For the first 15 years or so of its existence the Cassegrain on the 200-inch was hardly used. In subsequent large telescopes it has been without question the most important observing station.[9]

Figure 6.6   The cell and back supports of the AAT primary mirror

6.3 The tube and mirror cells

Once the JPC had accepted the optical design, the Technical Committee turned to the mirror support systems. In Chapter 2 we pointed out how a horizontal disc-like mirror supported around its circumference will sag under its own weight by what is usually a very significant extent. In the case of the AAT disc this sag would amount to 6.8 microns, or 13.5 wavelengths, a distance tiny by mechanical but large by optical standards (a wavelength is taken as half a micron). To produce a perfect image, the mirror should maintain its shape to about a tenth of a wave, regardless of the direction in which the telescope is pointing. This meant that the AAT mirror had to be counterbalanced to better than one part in a hundred, and the counterbalancing had to operate from both back and sides, or both axially and radially, in such a way that when the telescope moved away from the zenith the side supports would take over smoothly from those at the back. That mirrors should be supported in this way was clearly recognised by the builders of the great speculum mirror telescopes last century.[10]

The support of the primary mirror was an especially serious problem with the 200-inch. The ribbed structure which had been adopted to minimise the effects of temperature fluctuations had the effect of making the mirror very flexible, about nine times more so than the Kitt Peak or AAT mirrors. The 200-inch mirror has

Figure 6.7   The AAT primary mirror showing its side supports

the rigidity of a 60-inch mirror a little over one inch thick, the AAT mirror that of a 60-inch three and a half inches thick. To compensate for the gravitational forces on the 200-inch mirror with sufficient accuracy was a considerable task, achieved only with great difficulty. In particular it required that friction in the supports be reduced to a minimum, and a satisfactory solution was found only at a second attempt. It employed a series of 36 elaborate counterweights working in gimbals and evenly distributed around the back of the the mirror. Each counterweight was reputed to have cost as much as a Cadillac.

Three options were available for the back supports of the AAT mirror – mechanical levers, air bags or pneumatic pistons. The choice went to the pistons because of their very low friction and because they could be tested individually in the manufacturer's works at any orientation of the mirror cell. In the final design the mirror was supported from the back by 36 pads arranged in two circles, and from the sides by 24 mechanical counterweights acting through levers in a push-pull mode (see Figs. 6.6 and 6.7). The pistons of 33 of the back pads were operated by air pressure which could be varied as the cosine of the zenith distance, while the remaining three functioned as fixed locating points. The pads were arranged so that 12 were in an inner and 24 in an outer circle, the radii of the two circles being optimised by a finite element computer programme. Such programmes were then new, and this was their first use in the project. Their ability to calculate accurate stresses and distortions for

Figure 6.8   Herman Wehner washing the AAT mirror before aluminising

quite complex structures is now taken for granted, but at that time it opened new horizons and gave the AAT engineers an immense advantage over the designers of the 200-inch. In the AAT case the programme solution was right on target and the cell worked perfectly on the first assembly, with no need for further adjustment.

The secondary mirrors all function face downwards. Each is held in its cell and pulled upwards against three fixed locating points by a partial vacuum. A reduction in the air pressure to only five per cent below atmospheric is sufficient to hold these mirrors in place. The edge supports consist of thin tubes or girdles partially filled with mercury. They also serve as vacuum seals, and have the further important function of locating the mirrors accurately in their cells. This is an elegant solution, but there were lengthy delays in making tubes which were sufficiently flexible and strong, and at the same time would ensure that the supporting pressure was directed accurately towards the centre of gravity of the mirror. At one point the Board felt impelled to engage a second firm in case the first, Sir Howard Grubb, Parsons and Company, did not produce the girdles in time. The material finally selected was silicone rubber, reinforced with Terylene and 0.038 cm thick. The tube for the f/8 mirror was especially difficult, with delays long enough to cause anxiety. This mirror was 147 cm in diameter and weighed almost a tonne, and supporting it upside down in the middle of the telescope tube was no trivial matter.

In the Kitt Peak telescope all three secondary mirrors, together with the prime focus observing cage, were carried near the top-end of the tube, and the mirrors were flipped in and out of the beam as required. This design allowed rapid changeovers but was complex and heavy, and at the longer focal length of the AAT it was impracticable. Also, it would have been difficult to adapt for other arrangements, such as a special chopping f/15 mirror for infrared observations for which a demand arose quite soon. Instead, the AAT designers decided on three separate top-end assemblies stored on the floor of the dome. Each unit weighs about four tonnes. They are interchanged with the telescope held vertically, and, helped by a semi-automatic crane mounted inside the dome, a changeover takes about 35 minutes. The optics are focussed by motors which drive the appropriate top-end as a whole, up or down the tube. The scheme has been most successful. The drive is refreshingly smooth, free from the judder and lateral shifts so common in these mechanisms, and the top-ends relocate after changeovers with barely a detectable error.

A second AAT innovation was an automatic focussing device which took advantage of the fact that the focal lengths of zero-expansion optics do not change with temperature. Focus shifts are then caused only by the thermal expansion or contraction of the steel tube, and can be eliminated by monitoring the distance between primary and secondary mirrors with a permanently mounted invar tape. Invar is a nickel steel with a negligible coefficient of expansion. The top-end can then be driven so that this distance is maintained constant regardless of the temperature. The AAT was the first big telescope equipped in this way. The system has been a boon to observers, to whom the need to check focus usually comes as an annoying and time-consuming interruption.

Much thought was given to problems of optical alignment, an awkward job which astronomers do not like. If the optics are misaligned the image quality is degraded, the central images developing coma. And if the alignment changes as the telescope moves around the sky, the pointing accuracy deteriorates. In the past, only the first criterion was considered, but with telescopes like the AAT, designed for really high pointing accuracies, the second was the more important. This aspect of the problem was first pointed out in an analysis by John Pope.[11] To eliminate coma the secondary mirrors should be centred to within about a millimetre of the axis of the primary, and squared on to half an arc minute. To maintain pointing accuracy as the telescope is moved, the centring should not change more than a tenth of a millimetre, and the squaring on by more than an arc second or so.

These formidable tolerances, which impose such severe conditions on the rigidity of the tube, first became important in connection with the corrector lenses for the 200-inch. There they were met by the invention of the Serrurier truss, named after its designer and shown in Figure 6.9. In this well-known device the tube is allowed to bend, but only in such a way that the two mirrors deflect through equal distances,

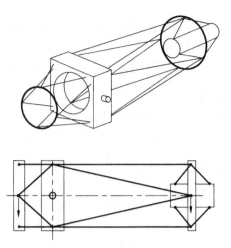

Figure 6.9 The telescope tube with Serrurier trusses. The trusses are designed so that when the tube is moved away from the vertical the two ends will bend through equal distances, remaining parallel while they do

at the same time remaining parallel. Collimation is thereby preserved regardless of the direction in which the tube is pointing. A Serrurier tube has been a standard component of almost every telescope since the 200-inch.

The intimate relation between the optical and engineering aspects of the telescope tube will now be evident, particularly as between the mirrors and their supporting cells. The desirability of having the mirrors and cells made by the same manufacturer will also be clear. The relevant tender documents for the AAT, therefore, were written so that the manufacturer could bid for either the tube and cells, or the optics, or both. The successful tenderer, Sir Howard Grubb, Parsons and Company, did bid for both. This well-known British company could trace back its origins to the third Earl of Rosse, whose family name was Parsons. With the engineer, Thomas Grubb, the Earl had built a famous 72-inch reflector, 'the Leviathan of Parsonstown', in the 1840s. Grubb Parsons was established by sons of these two men. It had been the source of most British and Australian telescopes and telescope technology for more than a century.

6.4 Accepting the primary blank

The JPC accepted the Owens–Illinois tender for the primary blank in November 1967. To allay any doubts about the long-term stability of the then novel material Cervit, the contract included a ten-year guarantee. A core sample was taken, analysed by X-ray diffraction and electron microscopy, then sealed back into the disc, to be remeasured if any question arose. None ever did.

A blank as large as 3.8 metres was a considerable step up for Owens–Illinois,

Figure 6.10   Owens–Illinois staff with 'the world's largest known single piece of glass' *(Photo: Owens–Illinois Inc.)*

whose previous biggest was 2 metres. The company had to build new furnaces, and re-examine the heat treatment and other procedures, and the AAT blank was not poured until 9 April 1969. When four weeks later it was taken out of the heat enclosure and trimmed up, it was hailed as 'the world's largest known single piece of glass'. It was 4 metres diameter, 71 cm thick and weighed 27.5 tonnes. Few sights in industry can be more impressive than pouring so large a blank in so short a space of time, and on its day it was a considerable media event.[12]

However, at about the same time the Russians poured a pyrex blank for their 6-metre telescope. It was more than twice as heavy as the AAT blank, so they took the title. The blank for CTIO, poured later than the AAT blank and the larger of the two, had to be described by the Owens–Illinois publicity department as 'the largest single piece of glass in the free world'.[13] The final diameter of the AAT blank was fixed at 3.94 metres, as nothing bigger could be fitted into the already designed tube. The clear aperture came out at 3.89 metres, the excess being needed for edging and chamfers.

At the end of its four weeks' heat treatment, the blank was ready for inspection by the Technical Committee. There were obvious potential problems such as the presence of sizeable bubbles which might break the mirror surface, but internal stress was the main concern. Stress usually arises when hot or molten material is cooled unevenly, especially if it is cooled quickly. The edge of a mirror blank cools and

Figure 6.11   The mirror for the CTIO telescope glowing white hot, poured just after the AAT mirror at the Owens–Illinois plant *(Photo: Owens–Illinois Inc.)*

solidifies first; when the interior cools it may contract (as it does for normal glass) or expand slightly (as it does for Cervit). The resultant internal stress can be relieved by long slow cooling known as annealing; the 200-inch blank was in the annealing oven for ten months. And a badly stressed mirror is anathema; its shape may change erratically over the years, and it may crack or even fracture without warning.

Because the expansion of Cervit was so low it was not expected to show much stress, but a proper check was nevertheless essential. Stress is normally detected with polarised light. The mirror is placed between a pair of polaroid sheets, correctly oriented, and any stress shows up as an easily recognisable pattern of light and dark areas. But while small pieces from the AAT blank were reasonably transparent, the full thickness was almost completely opaque. The first attempt by Owen–Illinois to resolve the impasse was to borrow a huge military searchlight and, as it were, blast the way through. So intense was the beam that a hand could be held in it only for a few seconds, and it made the blank too hot to touch; but it was invisible from the other side. These trials were made at night, in the huge shop which housed the Cervit furnaces. The intense blue glare of the searchlight, the subdued roar and deep red glow of the furnaces, the curious acrid smell, the milling people and the general air of disquiet, all combined to make the night unforgettable.

Next day, after much anxious discussion, the blank was found to be somewhat more transparent in deep red than in blue light, and this led to a solution. The blank thickness was to be reduced as far as the specifications allowed, test areas would be rough polished to transmit more light, and time-exposure photographs taken of them with emulsions sensitive in the deep red. There were complications, which it took three visits to rectify. However, David Brown, of Grubb Parsons, and Redman who carried out the final inspection found the stress comfortably within specification, and had no hesitation in accepting the blank.[14] The secondary blanks were tested and accepted at the same time. Being much thinner they were reasonably transparent, and there were no complications.

There was a final detail. In the account for the packing cases an unexpected item was found, $3500 for armour plate. The original plan had been to send the blanks by rail to a southern USA port and ship them to England from there. Apparently a case as big as that for the AAT primary mirror, travelling by train across the prairie, would have been an irresistible target to any wandering cowboy. As one of the packers explained, '*It's not the hand-guns and the 45s, it's the high-power rifles that worry us*'. In the event the blanks went by way of the St Lawrence Seaway and the North Atlantic, reaching Grubb Parsons at Newcastle–upon–Tyne at the end of 1969.

6.5 Specifying and testing the mirrors

How good does the final mirror need to be? If the irregularities in its polished surface, the hills, the undulations and the valleys, do not exceed about a tenth of a wavelength (there are 50,000 waves of blue light in an inch, 2000 in a millimetre), optical theory tells us that the mirror may be accounted perfect. But, because of an effect known as diffraction, a consequence of the wave nature of light, the image that even a perfect mirror makes of a point is not itself a point, but a tiny disc. It is called the diffraction disc, and its diameter is $5/D$ arc seconds, where D is the mirror diameter in inches. This diameter is the the resolving power, the smallest detail the mirror can distinguish. It is about 1/30 arc second for a 'perfect' 150-inch mirror, and such a mirror is said to be diffraction limited. Any defects on the mirror surface will smear the image and impair the resolution.

In practice, defects like this are often overwhelmed by seeing, a term already discussed in Chapter 5 and used to describe the extent to which stellar images are degraded by inhomogeneities or turbulence in the earth's atmosphere. Seeing is akin to the shimmer visible on a hot day, and in its presence a star-like point is imaged into a disc of a different sort, the so-called seeing disc, which may vary in size from less than a half to ten arc seconds, depending on the degree of turbulence. At any site sub-arc second seeing ranks as extremely good; a good site will average one to two arc seconds, while four is distinctly mediocre.

Clearly a mirror should be good enough to take advantage of the best seeing the site has to offer. In the late 1960s an excellent mirror was regarded as one which would put 50 per cent of the reflected light into a circle 0.3 arc seconds in diameter, and 80 per cent into a circle 0.6 arc seconds in diameter.[15] In fact, at that time, no better large mirror had been made. However, in 1967, when the optical specifications were being drawn up, several reasons suggested it would be worth trying to achieve better. The AAT mirror was to be solid and substantially stiffer than the ribbed blank of the 200-inch; it was to be made of zero-expansion Cervit, and therefore would not distort with changes in temperature. Finally, some very good seeing had been experienced at Siding Spring. Therefore, the specification for the primary mirror was made tighter than the above figures and written in two stages. In stage I 80 per cent of the light was to go into a circle of 0.4 arc seconds diameter, 95 per cent into 0.7 arc seconds, and 99 per cent into a 1.0 arc second diameter circle. If stage I were reached, there was an option to go to stage II, according to which the above diameters would become 0.3 arc sec, 0.5 arc seconds, and 0.8 arc seconds respectively.

This is the conventional language of the encircled-energy criterion. It is best interpreted as a condition on the slope of the error, where the error is the amount by which the actual surface departs from the desired surface anywhere on the mirror. If the slope of the optical surface is in error by 0.1 arc seconds, a ray reflected from the mirror at that point will be deflected by 0.2 arc seconds and will intersect the focal plane at a distance of 0.2 arc seconds from the focus. According to the stage I specification, the AAT mirror had to satisfy this condition over 80 per cent of its surface. A condition as stringent as this implies that the height error in the mirror surface will not exceed a fraction of a wavelength, in which case ray or geometrical optics are no longer applicable and the test results must be interpreted by different methods. The resulting images are in general much closer in size to the diffraction images described above.

Optical testing is an extensive subject and has its subtleties. Here we discuss it only briefly. Nevertheless, testing the AAT mirror was obviously important, and at times of much concern to both the Board and the Project Office. In normal telescope usage the beam incident on the primary mirror is parallel, but testing is almost invariably carried out at the centre of curvature. In the usual arrangement a point source, such as an illuminated pinhole, produces a diverging beam which the mirror converges to form an image close beside it. This can be examined in a variety of ways, the devices used ranging from a laser-powered, unequal-path interferometer to the simple knife-edge (or half razor-blade) used by amateurs and professionals alike for more than a century. Only for a sphere is the interpretation of the test results obvious, and for test purposes it is advantageous and sometimes essential to make an aspherical mirror simulate a sphere. This can be done with a device called a compensator, a common form of which, the Offner compensator, is the two-lens pair illustrated in Figure 6.12.

The testing of large mirrors is highly sensitive to environment: air movements, temperature gradients and any form of ground vibration, all of which must be eliminated as completely as possible. The ideal place to test a big mirror would in fact be in outer space, where there is no air, no vibration, and no gravity to distort it.

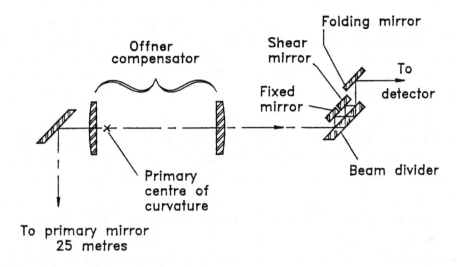

Figure 6.12 The Grubb Parsons test rig for the AAT primary mirror showing the Offner compensator and shearing interferometer

Failing this it is best to work from directly overhead. In a calm atmosphere air strata form horizontally, and do the least damage when the line of sight is vertical, crossing the strata perpendicularly. Also, for test purposes the best way to support a mirror is on its back, horizontally. Hence the high test towers so characteristic of large-component optical shops, with their double-walled test chambers, stringent air control and anti-vibration mounts.

One of the most commonly used test instruments is the unequal path interferometer. It measures directly the error on the the mirror surface and presents the result in the form of a contour map. The contours appear at every half wavelength because errors on the mirror are doubled on reflection. The test is ideal for the optician because it shows him exactly where to polish off his next layer of glass. But the method is sensitive to air movements along the line of sight, which make the fringes shift continuously. Repeated readings must be taken, and a rather complicated reduction procedure adopted to average out the variations. The test is also very sensitive to ground-transmitted traffic vibrations.[16]

A second method, known as the Hartmann test, measures the slope of the error on the mirror rather than the error itself, and so relates directly to the encircled-energy criterion above. First proposed in 1904, it uses the simple method illustrated in one-dimensional form in Figure 6.13, which shows the test as carried out in a telescope, using a star as light source. It has been familiar for decades and must have been used to test the optics of almost every big telescope in existence. With this method the mirror is covered by a screen pierced with a pattern of holes. The AAT screen had

432 holes, 5 cm in diameter and arranged on a 15-cm grid. The holes (a-d) define thickish bundles of rays (Fig. 6.13), and if the part of the mirror behind one of the holes is slightly tilted, as for example the shoulder of a low bump, the relevant ray will pass near but not through the focus. The measured offset indicates the slope error at that part of the mirror.

In practice a photographic plate is used slightly out of focus to record a pattern of dots. The dot positions can be measured accurately from the plate, and from the measurements one can infer where the focal plane lies, and the scatter of the rays about the focus itself. The smaller the scatter, the better the optical quality of the mirror.

A third method uses a wave-shearing interferometer (WSI) which measures differences in surface contour between adjacent points rather than actual errors, as with the interferometer, or errors in surface angle, as with the Hartmann test. It measures in fractions of a wavelength the difference in height between the mirror itself and a replica of the mirror displaced laterally by a short distance s. This distance, the shear, is introduced by an optical device. From the pattern of differences an error map similar to that from the laser interferometer can be constructed. In mathematical terms, if $E(x, y)$ is the error at the point (x, y), the laser interferometer measures $E(x, y)$, the Hartmann test measures dE/dx and dE/dy simultaneously, and the shearing interferometer measures $E(x + s, y) - E(x, y)$. The interferometer is then rotated through 90° and measures $E(x, y + s) - E(x, y)$ in a separate observation; s is the shear. The Grubb Parsons optician, David Brown, introduced the shearing interferometer into industry, and over the years developed it to the stage where it became the standard Grubb Parsons test instrument. It has the particular advantage that it is almost completely immune to the effects of air movement and ground vibration, a major consideration in the heavily industrialised area where the Grubb Parsons works were located.

The test rig used by Grubb Parsons for routine and final testing of the primary mirror is shown in Figure 6.12. It was located at the centre of curvature, 25.4 metres vertically above the mirror, where it was reached by ascending a very large number of stairs. The compensator was referred to above. The components of the shearing interferometer are also shown: mechanically it is a very simple device, but its mode of operation is too technical to be explained here.

For the final acceptance tests, reliance was placed primarily on the Hartmann method, with the results interpreted geometrically. This was an important decision, because the works tests had to be conclusive; the possibility of returning the mirror from the other end of the earth for further work could not be entertained. Although it was known to be laborious and time-consuming, the Hartmann test had advantages. It was accurate and straightforward and required a minimum of equipment; it

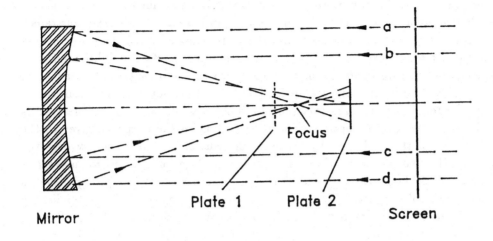

Figure 6.13 A simplified form of Hartmann test

avoided interpolation, as between interferometer fringes; no intermediate optics were required, especially no compensator; it was relatively unaffected by air conditions and vibration; finally, its results were quite independent of those from the shearing interferometer.

The final workshop verification of the quality of large mirrors was made possible by advances in test methods.[17] Previously, primary mirrors were not accepted until they had been tested in the telescope, and examples were not lacking of mirrors which had been found wanting and in need of further work. The Mount Stromlo 74-inch mirror had to be returned because it was astigmatic, and the outermost zone of the Palomar mirror came out, as planned, a little high. Don Hendrix, the optician, removed the excess with hand polishers in just nine hours, a legendary feat.

6.6 The finished mirror

Grubb Parsons received the AAT blanks at the end of 1969 and began work on them at once. By early 1971 specification I had been reached, and further tests showed that the primary mirror maintained its good performance when mounted in its cell, and that the characteristics of Cervit had remained extremely good. It therefore was decided to proceed to stage II. To reach this stage some slight astigmatism had to be polished out. It was a nerve-wracking process which required considerable skill, and it delayed progress on the mirror by several months. Such delays, however, are not unusual in this type of optics, and by March 1973 the mirror was ready for final testing.

The figuring of large mirrors is a deliberate process which has not changed much with the years. Amateur and professional opticians alike use much the same methods, and in fact some of the best-known professionals graduated from the amateur ranks. The mirror is mounted face upwards on a slowly rotating turntable, and a polishing tool of equal and opposite curvature, lined with pitch facets and lubricated with a suspension of cerium oxide, is moved to and fro across its surface. Because the curvature of a hyperboloid (or paraboloid) is not constant but decreases from centre to edge, the tool usually works over only limited areas of the mirror. These areas are often zones, which the optician selects according to the shape he gives to the polishing area of the tool, and by the path he causes the tool to follow over the mirror. At the end of a run, of say half an hour's duration, the mirror is washed down and tested. The optician decides on the next areas to work on, the polisher is reformed, the stroke adjusted if necessary, and the cycle repeated. Not all polishing is done by machine – isolated high areas are sometimes reduced by hand. Finally, the sensitivity of the image to any hint of ellipticity or astigmatism in the mirror is most important. Because circular symmetry must be maintained at all costs, extreme care is taken with support systems and with the regular rotation of mirrors on their supports throughout the polishing stage. Progress is necessarily slow. In its final stages a mirror may be given only two polishing runs a week, the rate of progress being governed almost entirely by the time taken to make the tests and to analyse the results.

There were two main series of tests for the primary mirror, the first made with the interferometer under the direction of David Brown, the second with the Hartmann screen, under Roderick Willstrop. Gascoigne and Redman represented the AAT Board in planning the tests and assessing the results. The interferometer tests were made with shears of 9 cm and 7 cm, the Hartmann tests with 5-cm holes, plus a later run with 2.5-cm holes. Extreme care was taken with the mirror support. During tests it was mounted on a water-bag and between runs it was rotated 90°. There were four exposures each run, and the average over four runs was taken as the result of a complete test. This procedure was adopted to reduce the effects of asymmetries in the support system, and of random errors caused by air movements.

In all, 20 Hartmann plates were taken, each with about 438 spots. They were measured on the automatic GALAXY machines at the Edinburgh and Greenwich Observatories. It was fortunate that these machines were available, because to have measured so many plates by hand would have been much less accurate and unacceptably time-consuming and tedious. The resulting body of data, with the position of each spot known to better than a micron, was surely the most accurate and complete yet assembled for a Hartmann test. It was worked over by a number of groups, including the Itek Corporation and the Arizona Optical Center.

Table 6.1 shows the direct results obtained within the AAT Project Office. X, Y and Z are the diameters in arc seconds of the circles which enclose 80, 95, and 99 per

Table 6.1. *Test results on the AAT primary mirror*

Test	Spacing cm	80% X	95% Y	99% Z	Q
Specification I		0.4	0.7	1.0	1.961
Specification II		0.3	0.5	0.8	1.119
WSI	9.1	0.221	0.333	0.476	0.494
WSI	7.1	0.253	0.490	0.644	0.853
Hartmann	5.1	0.337	0.496	0.651	1.065
Hartmann	2.5	0.390	0.609	0.825	1.555
Mean		0.300	0.482	0.649	0.951

cent of the reflected energy. The quantity Q is a figure of merit needed for contractual purposes and to describe the mirror performance. $Q = 4X^2 + 1.33Y^2 + 0.67Z^2$, and was effectively a measure of the area of the image of a point source. It will be recalled that specification I was regarded as the minimum figure acceptable, specification II as a target to be attempted after specification I had been reached. Averaged over the four tests, the final value of Q was 0.951, and placed the mirror formally within specification II.[18]

But it was disconcerting to say the least, and totally unexpected, that the quality of the mirror should depend so obviously and so regularly on the test spacing. It was this very regularity, however, which pointed to the explanation. With the aid of other lines of evidence, it was finally concluded that the row-to-row variations were due to irregularities only a few hundredths of a wavelength high. Such amplitudes were too small to have any effect on the diffraction image, but they would introduce into the Hartmann results random additional slopes which would be greatest for the smallest holes. The effect would smooth out increasingly as one tested with larger holes or shears.[19]

The data can be presented in many ways, and were in fact discussed at great length. Figure 6.14 is a picture of a knife-edge test of the traditional kind. It shows small systematic errors in the form of 'tree-rings' and 'spokes', but generally reveals a satisfactorily smooth figure. By integrating the Hartmann results Willstrop obtained a contour map of the errors on the mirror surface which yielded an average (r.m.s.)

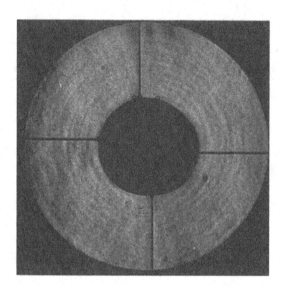

Figure 6.14   A knife-edge picture of the primary mirror taken in the telescope through a corrector. The circular marks ('tree-rings') were made during the normal figuring process, the radial marks ('bicycle spokes') when removing a small amount of astigmatism. They correspond to features which are not more than a few hundredths of a wavelength high or low. The three plumes at the top were made by jets of warm air entering the beams from holes in the centre section.

error of 0.13 wavelengths. This would have been a creditable result for a 25-cm optical flat, and suggested that the mirror was near diffraction-limited. This suggestion was born out by diffraction images (point-spread functions) calculated by the Itek Corporation (Fig. 6.15) and by Willstrop. What carried instant conviction to most people, especially to the astronomers, were the photographs of artificial double stars in Figure 6.16 where the mirror easily resolved an equal brightness pair with a separation of only 0.17 arc seconds. It was perverse that this, something of an afterthought, should have been the conclusive test, while the Hartmann test, on which so much reliance was placed and so much labour spent, was made to yield an answer only with difficulty.

The exercise had certainly achieved its primary objective, the production of first-class optics for the AAT. The AAT primary was the first large mirror for which diffraction had to be taken seriously, and it was the AAT's success which encouraged KPNO to persevere until the mirror for its CTIO telescope had achieved an equivalent high standard. An unexpected result was to show that when used for testing mirrors of the highest quality, the Hartmann test had its shortcomings. This was certainly the view taken with the 4.2-metre mirror for the William Herschel Telescope, also figured by David Brown. The specification for this mirror was written mainly in terms of an auto-correlation function of the surface errors; this is easily interpreted

Point spread function

at $\lambda = 3250$Å   at $\lambda = 5780$Å

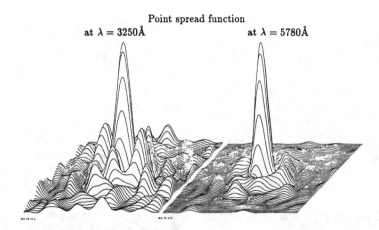

Figure 6.15 Image profiles (point-spread functions) computed by the Itek Corporation from the final Hartmann tests of the AAT primary mirror. On the right, the central intensity is about 0.5 of the aberration-free intensity (Strehle ratio 0.5).

in terms of the measurements made by a wavefront shearing interferometer, and it is also directly related via diffraction theory to the modulation transfer function of the mirror.

Packed in a steel case painted bright red, the AAT mirror was shipped out of London for Sydney late in the autumn of 1973. It travelled the 500 km from Sydney to Coonabarabran by semi-trailer, under police escort on a route chosen to avoid various narrow bridges and low underpasses. When it arrived on 5 December, it was met by the Shire President, the Council, assembled citizenry and a request to do an additional circuit of the town so that everyone else could see it. *'Like a lap of honour after winning the Bathurst Five Hundred,'* said the driver. (The Bathurst Five Hundred was the premier motorcar race in Australia at the time.)

6.7 The secondary mirrors

The Grubb Parsons contract included the figuring of the three convex secondary mirrors for the f/8, f/15 and f/36 (coudé) foci respectively, and of the three coudé flats. Table 6.2 contains physical data on the mirrors, together with the costs of the blanks and of the figuring.

Note how it cost almost as much to figure the secondaries and flats as to figure the primary mirror. This was partly because the secondaries were all convex and therefore more difficult to test. Secondaries are almost invariably hyperboloids, with one focus inaccessible, and because of this they are usually tested with the aid of an auxiliary test mirror, called a Hindle sphere. To test the f/8 secondary in this way would have required a sphere of about 2.5 metres diameter, a major piece of optics which it would have been necessary to figure and mount very carefully. Grubb Parsons used instead a meniscus lens. It was appreciably smaller, although it had

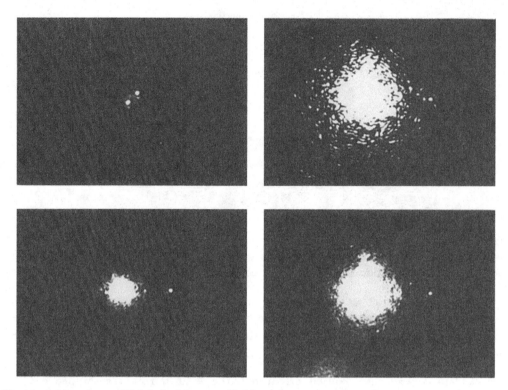

Figure 6.16   An artificial double star (left) with a separation of 0.17 arc seconds and intensity ratio of 1 to 1, and (right) with a separation of 0.63 arc seconds and an intensity ratio of 100 to 1 *(Photo: Sir Howard Grubb, Parsons and Company Limited)*

the drawback that because its central section was traversed by the test beam twice it introduced aberration which had to be removed by further optics. Generally, however, this latter method was cheaper and better.

Another difficult piece was the No 1 coudé flat, the oval diagonal mirror which directed the beam along the declination axis. John Anderson, who was in charge of the 200-inch optics, said of its coudé flat, that on account of its shape and relatively large size (91 by 135 cm), it was in many respects more difficult to make than any of the other mirrors in the telescope. However, David Brown, who was responsible for the AAT optics, said of the corresponding AAT flat which was 71 by 107 cm only that it gave more trouble than any other of the smaller mirrors.[20]

The prime focus correctors also had to be made, and after tenders had been invited internationally this contract went to William E. James, a maker of precision optical surfaces in Melbourne who had made his name in Australian scientific and academic circles. With a biggest lens of 46 cm diameter, the AAT triplet corrector was the largest to have been made up to that time.[21]

Table 6.2.   *Physical data about the AAT mirrors*

(a)   Costs in $A

Blank for the primary	541,000
Blanks for the secondaries and flats	58,000
Figuring the primary to stage I	204,300
Figuring the secondaries	144,900
Figuring the flats	58,500

(b)   Mirror dimensions

Mirror	f/no	Focal length	Eccentricity	Diameter	
		m	e	cm	inches
Primary	3.26	12.700	1.0825	389	153.3
f/8 secondary	7.87	−7.249	2.8825	141	55.3
f/15 secondary	14.80	−3.361	2.0465	84	33.1
f/36 secondary	35.84	−2.888	1.5944	83	32.6
No 1 coudé flat		∞		107 × 71	42 × 28
No 2 coudé flat		∞		71	28.0
No 3 coudé flat		∞		34	13.0

No 1 coudé flat was elliptical, and the figures quoted are the longest and shortest diameters

6.8 A master optician

David Brown, who contributed so much to the success of the AAT and the UK Schmidt Telescope, and whose name appears so often in this book, died in 1987 while it was being written. As was acknowledged throughout the astronomical world, he was a past master in the art of figuring large elements for telescopes, with hardly an equal in his generation. He must certainly be ranked with Ritchey and other great opticians of the past, and is assured of a high place in the hierarchy of British instrumental astronomers. Other major achievements include the optics for the 2.5-metre Isaac Newton Telescope, the 3.5-metre UK Infrared Telescope, and the 4.2-metre William Herschel Telescope. Besides his scientific abilities he was a man capable of warm friendships, and his long association with Gascoigne and Redman had the character much more of a personal collaboration than a business relationship.

Notes to Chapter 6

1 di Cicco D. (1986). *Sky and Telescope* **71**, 347
2 Redman R.O., internal report on his 1967 American visit
3 The primary mirror of the Cambridge 36-inch, made of Pilkington low-expansion pyrex, had partially devitrified while being annealed. Since then it had deteriorated further, in the process becoming steadily more difficult to aluminise. Redman had good reason to be cautious about partially devitrified glassy materials. (Letter from Willstrop to Gascoigne.)
4 The first was the Kitt Peak 150-inch mirror, made of General Electric's fused natural quartz.
5 The maximum departure of a conic mirror from the sphere which contacts it at centre and edge is $e^2 D/(16F)^3$; e is the eccentricity of the mirror, D its diameter and F its focal ratio. The Schwarzschild constant b is equal to $-e^2$. For the AAT mirror this departure was 64λ or 32.2 microns, for the KPNO mirrors about 116λ.
6 The KPNO, CTIO and AAT mirrors were all planned to be 3.8 metres in diameter. An additional layer of quartz had to be added to the first KPNO blank, and in the process the heated blank flowed enough to make a 4-metre mirror possible, though only the inner 3.8 metres are actually used. The CTIO (Cervit) blank was poured to 4 metres, and the AAT blank came out oversize enough to allow the clear aperture of the mirror to be 3.89 metres. The overall aperture is 3.94 metres.
7 Ross's original two-lens correctors were designed for the Mount Wilson 60- and 100-inch telescopes, both f/5 (Ross F.E. (1935) *Astrophys. J.* **81**, 156; Wynne C.G. (1965) *Applied Optics* **4**, 1185). He had to add a third lens to the corrector for the f/3.3 200-inch mirror. A great deal hung on this corrector, and Ross worked under extreme pressure while designing it.
8 CCD is a charge-coupled device, a highly efficient solid state light detector, especially in the infrared and near infrared, and one of the essential devices of modern astronomy. For a short non-technical account see Henbest N. and Marten M. (1983) *The New Astronomy*, Cambridge University Press, pp.51–53
9 cf. Gascoigne S.C.B. (1970). *J. Phys. E.* **3**, 167
10 The need for good support systems must have been made evident by the properties of speculum. Besides depending on its diameter and thickness, a mirror will bend directly as the density of its material, inversely as its Young's modulus of elasticity. Speculum metal is more than three times as dense as glass or quartz-like materials, but has about the same Young's modulus. Therefore, a speculum mirror will bend three times as much as a glass one of the same dimensions because it will be three times as heavy. If we consider two mirrors of equal diameter and weight, the one of speculum will bend about 30 times as much as one of glass (it will have less than a third of the thickness).
11 Pope J.D. (1971). 'Optical Performance Criteria for Telescope Tube Design' in ESO/CERN Conference on Large Telescope Design, ed. R.M. West p.299
12 It was featured on the front cover of *Sky and Telescope* **38**, September 1969. See also pp.140–143, same issue, for an interesting account of Cervit and its manipulation.
13 Press release by Owens–Illinois Inc. on 25 June 1969
14 The complications were at least partly due to the considerable Rayleigh scattering produced by microcrystalline structure in the Cervit. While this was the reason

for the better transmission at longer wavelengths, it also raised the question, was the absence of structure in the test photographs due to depolarisation by multiple scattering or to the hoped-for absence of stress birefringence? Brown and Redman had to do a good deal of experimenting and calibrating before they could convince themselves that it was the latter which was true, and that the internal stress was indeed below the acceptance level of 10 nm/cm. (Letter from Brown to Gascoigne, June 1987)

15 Meinel A.B. (1969). *Applied Optics and Optical Engineering* – V, ed. R. Kingslake, Academic Press, p.136

16 The Arizona Optical Center is located about 500 metres from the University football stadium. Although the optical testing area is four floors below ground level, it is unusable on the days of a big match; testing cannot be resumed until the early hours of the following morning. (Personal communication between Wehner and Gascoigne).

17 Jeffery foresaw this clearly. In a covering letter to bidders for the tube/optics contract, he said: 'It is our belief that the primary mirror and support system can be shipped to site with an absolute certainty of their performance, following the test procedures laid down in the Specifications.' (RGO 37/467, 31 January 1969.)

18 If the final Q was 1.119, Grubb Parsons would receive the full sum for the mirror, if it was 1.961, they would receive two-thirds. For intermediate values they would be paid pro rata. Brown and Redman devised this scheme.

19 The micro-error explanation was first proposed by Gascoigne in an internal report. David Brown then produced the following neat example: 'If the only error on the mirror was sinusoidal with a crest-to-crest wavelength of 3 inches and a crest-to-valley amplitude of $\lambda/20$, it would, in an encircled energy test and using a geometrical calculation, fill almost uniformly a circle of diameter 0.4 arc seconds. The r.m.s. wavefront error of this defect is 0.016λ and the resulting diffraction maximum would be depressed by only one per cent (Strehl ratio 0.99), so that for practical purposes the error is negligible. The removal of fine structure of this amount from aspheric surfaces is very expensive.' [Brown D.S. (1973) *Observatory* **93**, 208]

20 It had 'a small local defect . . . comfortably within specification . . . nevertheless considerably larger than any other known error in any of the AAT optical surfaces' (Letter from Brown to Gascoigne).

21 Opticians can emerge from unexpected backgrounds. James was originally a professional actor with J.C. Williamson, a well-known Australian company in the performing arts. Finding himself increasingly given to polishing mirrors in his off-stage periods, he finally abandoned acting to take up a full-time career manufacturing precision optics.

7 Mounting, drive and control

7.1 The Kitt Peak mounting

In retrospect, and in the light of the way events actually unfolded, it seems curious that the AAT should have been so explicitly required to follow the design of the Kitt Peak telescope. In some quarters this was taken as a manoeuvre to discourage any move to build an alt-azimuth mounting, support for which had been growing; other people saw it as forestalling the sort of disagreements which had plagued the Isaac Newton Telescope.[1] In 1967, however, it was widely accepted that only the Americans knew how to build big telescopes, and it seemed the obvious way to go. Redman was one of the few who did not share this view.[2] The Kitt Peak engineers had already spent four years on their design, which was therefore well advanced, and its adoption could clearly lead to considerable savings in time, money and effort. It was only gradually that the AAT engineers realised that they too had something to offer, and during the course of the project they did indeed make advances which have had a considerable influence on the art of building large telescopes.

The idea that the AAT should use the Kitt Peak design was first mooted in 1962, almost before work at KPNO had begun on it. In 1963 Bart Bok raised the proposal in a letter to Sir Richard Woolley.[3] He and Alex Rodgers, both from Mount Stromlo Observatory, had just been to the west coast of the USA, where they visited the major observatories and talked to many of the leading figures in the large telescope world. KPNO, generous as always in such matters, offered its full co-operation, and finally, at the instigation of the Australians, the inclusion of the Kitt Peak design was stipulated in the interim Agreement for the AAT. Two KPNO representatives, David Crawford, Project Manager, and James Miller, Administrator, attended the April 1967 discussions in London where arrangements about sharing of drawings and other information were settled.

As described in Chapter 2, the AAT and KPNO technical groups first met as such in Tucson in May 1967, at a meeting called by the AURA Board to finalise the design of its 150-inch telescope. The proposed design was adopted and put out to tender, and before the end of the year AURA had entered a contract with the Western Gear Corporation of California to manufacture two complete mountings, one for Kitt Peak itself, the other for KPNO's Cerro Tololo station in Chile. The Kitt Peak group would have been happy to have the AAT go along with it, but while satisfied that the basic design was sound, the AAT engineers had reservations about certain aspects of it and were unwilling to commit themselves at such short notice.

The Kitt Peak design stemmed directly from that of the Palomar 200-inch. It was based on a study carried out by the Westinghouse Electric and Manufacturing Company, which 30 years before had been a major contractor for the 200-inch mounting. The main difference arose from the need to stiffen the horseshoe which forms the main support for the telescope tube (Fig. 7.1). To do this, it was proposed to move the declination axis from its position midway between the horseshoe and the lower (south) bearing, and to locate it in the plane of the horseshoe, where it could be used as a stiffening member. Some stiffening was clearly desirable, the new arrangement was more compact, and economies in layout gave a further advantage. The Palomar coudé arrangement was also modified, and a train of five mirrors adopted in its place.

One early problem has been discussed in Chapter 6. The increase in the focal length of the AAT primary mirror meant that the tube had to be lengthened, and this ruled out Kitt Peak's rather elaborate top-end assembly for interchanging secondary mirrors – it would now have been too long and heavy. In any case, the AAT group preferred separate interchangeable top-ends, even at the cost of the longer time it then took to switch between focal configurations.

This was not a major change, but a more deep-seated problem now arose. The horseshoe mounting has advantages which have led to its adoption for almost all the large equatorial telescopes, though it works less well at high latitudes. Variants of this mounting are shown in Figure 7.1. Its symmetry enables the tube to be supported from both sides, it gives access to the region of the sky about the celestial pole, and it leads naturally to the use of low-friction oilpad bearings. However, like all mountings it has its problems. The most important is that it is not an inherently stiff structure. At extreme hour angles the horseshoe can bend so much it no longer fits the oilpads properly. In the 200-inch this problem was met by deforming the horseshoe during machining in such a way that when the telescope was far east or far west, the horseshoe bent back into the shape of a true circular cylinder. Further, the horseshoe can twist and the tube centre section deform in such a way that the declination bearings can be pulled out of line by 30 or 40 arc seconds. These bearings are a critical component of large equatorial telescopes. They carry the tube and optics, a load which with the AAT exceeds 100 tonnes, and which must be able to move smoothly and freely at all times. Misalignment of the bearings can make them stiffer, can create stick-slip friction which is disastrous for the pointing accuracy and the smoothness of slow guiding motions, and in extreme cases can cause binding in the bearing itself. In the 200-inch the declination bearings were mounted in a bicycle-spoke arrangement which allowed them to keep in line, but there was not room for this in the Kitt Peak-AAT design. The bigger the telescope, the worse these problems become.

To help keep the horseshoe in shape, the Kitt Peak designers proposed to apply a pre-load of 50-odd tonnes to the declination axis so that it pulled the horseshoe arms together, and kept the declination axis in tension whatever the hour angle.

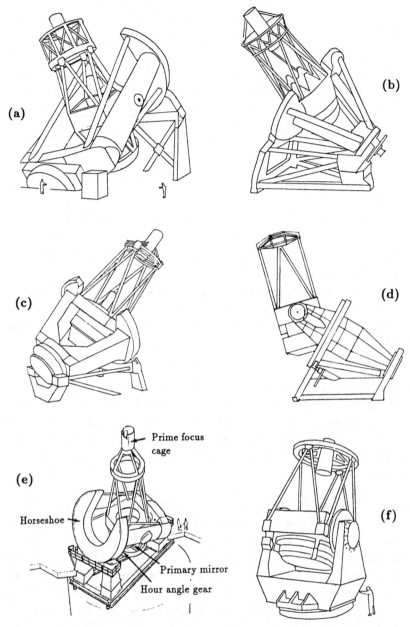

Figure 7.1 Simplified diagrams of some large telescopes chosen to emphasise particular features. (a) the Palomar 200-inch: note the position of the declination axis; (b) the Kitt Peak Telescope; (c) the Anglo-Australian Telescope: note the massive box struts connecting the horseshoe to the lower journal; (d) the ESO Telescope: the declination axis is now carried on a short fork, and the load of the tube counterweighted on the other side of the horseshoe; (e) the Canada–France–Hawaii Telescope: the declination axis is located as in the Palomar telescope, while the hour angle gear is attached to the horseshoe; (f) the ESO New Technology Telescope and its alt-azimuth mounting.

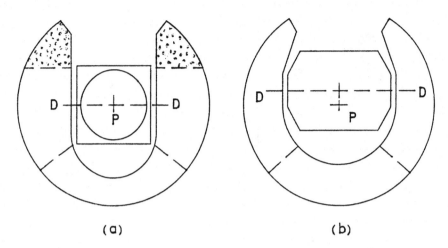

Figure 7.2   The (a) Kitt Peak and (b) AAT horseshoes. The AAT horseshoe differs by
(i) its wider throat and more uniform cross-section; (ii) the 107-cm separation between
the polar axis P and the declination axis DD; (iii) the more massive centre-piece of the
declination axis (this was the heaviest single component in the AAT); (iv) the elimination
of the concrete counterweights indicated in the tips of the KPNO tines. Both horseshoes
have diameters of 12 metres.

However, such a high preload meant a considerable increase in the internal friction
of the declination bearings. The resulting frictional torque was calculated as 5000
pounds feet, as opposed to 30 pounds feet for the right ascension bearings, and doubts
arose as to whether a motor could be found powerful enough to drive the telescope, let
alone give it a smooth trouble-free motion. These problems were clearly recognised
by Pope and Wehner.

   In January 1968 M.H. Jeffery took over as Project Manager, at just the time when
his expertise in structural engineering was most needed. The mounting problems were
tackled at once, and when the JPC next met, in March 1968, the Project Office was
ready to recommend that the horseshoe be redesigned by offsetting the declination
axis upwards from the polar axis by about one metre (finally 107 cm), and by making
the horseshoe centre section thicker and the side arms thinner. This was expected
to make it five or ten times stiffer; at the same time the 25 tonnes of concrete
counterweight carried in the tips of the horseshoe would no longer be required to
balance the structure. In addition, below the declination axis the horseshoe would
have a near-constant cross-section, which would have structural advantages. The two
designs are compared in Figure 7.2.

   The JPC approved the proposal and the Project Office went ahead with a preliminary design. Though hardly foreseen as such, this was a noteworthy decision in that
it marked the first step towards a complete redesign of the whole polar axis. Initially

the work was carried out in conjunction with the Canadian firm of Dilworth, Secord and Meagher, then closely associated with the ill-fated Canadian QEII telescope.[4] This firm supplied a number of the ideas incorporated in the final design.

7.2 The Ford–Minnett Report

In order to track on a star, a telescope needs to be driven about the polar axis almost at the same rate as the hour hand on a 24-hour clock, a rotational speed of 15 arc seconds per second of time. This speed must be maintained accurately and must be exceedingly smooth and steady. Enough variation should be available to compensate for the effects of refraction by the earth's atmosphere and for various maladjustments and flexures of the telescope, and to enable tracking on objects like the moon, which moves about three per cent slower than the stars.

Other rates are also necessary. To slew rapidly from one object to the next, large telescopes are usually driven at about 45° per minute in each co-ordinate. In the past it was customary to use separate motors and gear trains for this purpose, the power required to accelerate either axis quickly up to slew speed being much greater than that needed merely to drive it against small frictional and out-of-balance torques at diurnal rate. In addition, setting speeds are needed for moving an object in the viewing field on to a spectrograph slit or similar device, and guiding speeds are needed, in both declination and right ascension, for taking out the small variations which occur in telescope tracking. Typical speeds for these motions are between 60 and 1 or 2 arc seconds per second of time respectively.

Early in the century the motive power for most telescope drives came from falling weights, with the setting and guiding speeds supplied by various forms of differential gearing, some of them most ingenious. Electric power steadily took over, less quickly or completely than might have been expected, and by the the mid-1960s it was clear that the drive and control systems of most optical telescopes were lagging well behind corresponding systems in industrial and military technology. The point was emphasised by the success of the new radio telescopes such as the Parkes telescope (1962) and the Jodrell Bank Mark II (1966), where computer control and servo drives were used with conspicuous success.

Here we should explain that a servo drive system is one in which the actual motion of a telescope is continually compared with its desired motion, in this case supplied by a computer; any deviation is amplified and fed back to the drive motors so that the deviation is corrected. By this means a relatively small command signal can control precisely the behaviour of a very large telescope. The Parkes telescope, a huge 64-metre alt-azimuth instrument, was a particularly good example. Here the telescope was made to follow the motion of a small and very precise master optical telescope. The master moved on a conventional equatorial mounting, so that the large radio telescope followed the correct equatorial movement even though it was

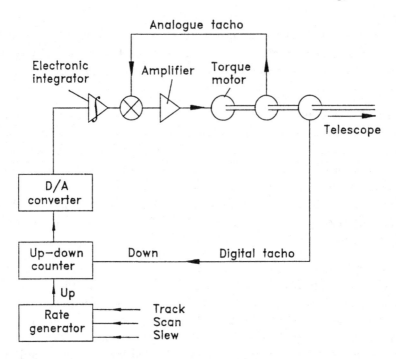

Figure 7.3 The Trumbo digital servo drive system. The drive motor together with a digital tachometer are mounted directly on the worm shaft. The tachometer is a shaft encoder which produces pulses at a rate proportional to the shaft speed, while the rate generator produces pulses at a rate proportional to the desired telescope speed. These are fed to the counter in opposite senses, so that the difference is always proportional to the shaft error. The difference is converted to an analogue signal which is amplified, then used to drive the torque motor. The analogue tacho loop gives control over the damping.

on an alt-azimuth mounting. At the time of its construction in 1960, the Parkes telescope was the most massive system under precise servo control.[5]

Among optical telescopes the shining exception to traditional technology was the digital servo system designed by Don Trumbo of KPNO, initially used for its 84-inch telescope, but since then widely adopted. His original system is sketched in Figure 7.3.[6] The rate of telescope movement on either axis is determined by a pulse rate. The basic rate is set by a rate generator, but it can be varied by adding into it or subtracting from it additional pulses, as required to correct, for instance, irregularities in the drive gearing. Normally this basic rate would correspond to the right ascension drive speed, but the system is accurate and flexible, and can follow any form of input, even a highly variable one. A spectacular demonstration of this is shown later. Such a system can be used for both polar and declination axes. It can handle a wide range of speeds, and because it uses digital techniques is well suited to computer control and to such devices as autoguiders.

When it came to drive and control there was no doubt where the JPC and the Technical Committee stood – firmly on the side of new technology. Late in 1967, H.C. Minnett from the CSIRO Division of Radiophysics, and L.E. Ford from the British Royal Radar Establishment, Malvern, were appointed as consultants to advise on these matters. Working rapidly, they visited key centres in Britain and the USA and produced a report in time for the JPC meeting in March 1968. Their principal recommendations were:

1. to incorporate *from the outset* an on-line digital computer, so that modern electronic technology could be used to the full to maximise the operational efficiency of the telescope, and to enable a high degree of automation to be achieved, especially through automatic guiding.

2. to adopt digital servo drives embodying the main features of the Trumbo drive system developed at KPNO, with attention to certain difficulties which had arisen at Kitt Peak.

3. to use semi-conductor logic as at Kitt Peak for the ancillary control system, together with KPNO's method of dealing with the problem of electrical interference by using a screened junction room.

Only Woolley, an unabashed conservative in these matters, opposed the recommendations. He made a stand for the old verities, complete with falling weights, but the radio telescopes had shown the way, and it was no contest.

7.3 Spur or worm gears?

The Ford–Minnett report had achieved its major objectives, but there had scarcely been time during its preparation to consider a number of servo control matters which had proved crucial during the feasibility studies for the Parkes telescope. To strengthen the design team in this and other areas the JPC appointed Jack Rothwell, an electrical engineer who had had considerable experience first with anti-aircraft gun control systems, then with the detailed design of the Parkes servo drive hardware. Subsequently he had been a member of the staff at the Parkes observatory and at the NASA Deep Space Station, Tidbinbilla, near Canberra.

A servo system must be precise and stable, qualities which depend not only on the design of the control system, but also on the characteristics of the mechanical and structural elements of the telescope itself, such as gearing stiffness, backlash, stick-slip friction, hysteresis and structural resonances. There were two immediate problems. One, which we defer until the next section, concerned the dynamical behaviour of the either axis structure (torsional resonances). The other was that servo engineers did not like worm gears in servo loops. They represented a high-friction, non-reversible element in the loop, and were prone to cause servo instability. In addition, worm gears could bind in case of sudden stoppage or reversal, and were difficult to provide with the efficient pre-load or anti-backlash devices essential in a servo system.

Jeffery and Minnett were attracted by the potential of spur (or helical) gearing, familiar to them from their experience with radio telescopes. Spurs provided a simple, reversible drive system, well adapted to servo control and to an elegant, symmetrical, anti-backlash arrangement using two drive motors. Much higher gear accuracy would be needed for an optical telescope however, because minute inaccuracies in tooth profiles might degrade the smoothness of the drive during the transfer of the load from one tooth to the next. It was essential that the fluctuations in telescope movement due to this factor be below those caused by atmospheric irregularities under the best seeing conditions.

Great advances had been made in the precision with which spur gears could be ground, but a change from the traditional worm gears would represent a major change in telescope engineering. A brisk controversy developed, inside and outside the Project Office. An alternative to gearing was studied. This was to drive the polar axis by friction, applied through a roller which bore against the horseshoe outer face. Such a roller could be machined with extreme accuracy and driven directly from a torque motor. If a horseshoe were driven in this way rather than by a torque applied at the 'little end', the effect of the torsional resonance referred to above would be largely eliminated. This was the course taken with the 3.6-metre Canada–France–Hawaii Telescope, which is driven though a 10-metre diameter gear attached to the horseshoe [Fig. 7.1(e)]. For the AAT, however, there were problems with encoding the polar axis, and a friction drive was a substantially new idea, with novel and untried features. Spur gears on the other hand were well tried on radio telescopes and well understood, and it had been established that they could be ground to the necessary precision.

So spurs it was, though this point was reached only after recourse to studies by yet more consultants. Once spur gears had been adopted there was no going back, and as Minnett put it, there was a good deal of anxious soul-searching before the final step was taken. The end result was a complete vindication of spur gearing. Moreover, the AAT's lead was followed by other large telescope groups, first by KPNO, which on the basis of its own investigations and of its access to the AAT reports, scrapped the worm gears which had already been made in favour of helical gears.

In the finally adopted right ascension drive the main gear is driven by two identical printed circuit motors working through identical gear trains. These motors have the advantage of smooth low-speed operation, which enables them to cover the whole speed range from guiding to slewing. They work in opposite directions, with the two gear trains connected only by a spring-loaded idler pinion in a well-known anti-backlash arrangement. The encoding is done quite separately, through a similar gear train driven from a different part of the main gear. They differ in that while the drive gear trains must be able to transmit enough power to accelerate their respective axes to full slewing speed within 12 seconds, the load on the encoder train is virtually zero. Thus, while the telescope is driven by the motors, the encoders are driven by the telescope. Provision is made for three separate encoders. One, the absolute encoder, reads to one arc second. The second reads to 0.05 arc seconds; this is the incremental encoder which is more sensitive to angular movement but which is less accurately calibrated. The third is the analogue or synchro encoder, intended as a backup in case of computer breakdown. The absolute encoders tell the computer where the telescope is pointing, the incremental encoders how fast it is moving.

However technical this may sound, the system was simple in concept, flexible and direct. Further, it was one of the fundamental concepts essential to the success of the telescope. The journey from concept to actuality was not short, and it took the dedication, hard work and expertise of many people to achieve it. The gearboxes were designed by Marconi Radar Systems Limited [an associated company of GEC-AEI (Electronics) Limited], and made by Mitsubishi as part of the drive and control contract, discussed below. The critical main gears each with 600 teeth and diameters of 3.66 metres came from Maag Gear-wheel, a specialist Swiss company, which was able to grind the teeth to the extreme precision, amounting to a few microns, required by a very exacting and detailed specification. The specification was described by W.A. Goodsell, the then Project Manager, as a product of *'blood, sweat, and tears – Jack Rothwell's blood, my sweat, and Harry Minnett's tears'*.

7.4 The polar axis and declination bearings

Shortly after the JPC meeting in March 1968, it occurred to Jeffery and Minnett that the torsional oscillations of the polar axis – the to-and-fro rotation of the horseshoe relative to the lower or north bearing – might interfere seriously with the functioning of the right ascension servo system.[7] The influence of similar torsional oscillations had been a critical design factor with the Parkes telescope. Calculations by Dilworth, Secord and Meagher predicted a frequency of 2 Hz, an uncomfortably low figure, and it was decided to look at means of stiffening that part of the mounting. Discussions with Dilworth, Secord and Meagher, supplemented by model tests made at the University of Sydney, continued for about 18 months before an acceptable solution was found. The KPNO telescopes used three tubular struts on either side to connect the horseshoe to the lower end journal. The decision was to replace these with the massive box girders which have become a distinctive feature of the AAT structure, and are illustrated in Figure 7.4. This change completed the divorce from the Kitt Peak design.

The problems with the declination bearings have already been described. The

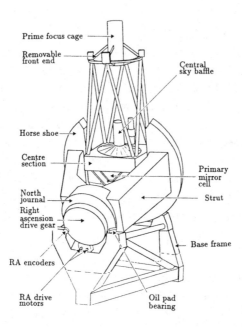

Figure 7.4   A sketch of the AAT as finally designed and built

redesign had made the polar axis much stiffer and enabled the preload to be reduced correspondingly. Nevertheless the declination bearings continued to provoke anxiety. This problem too was discussed with KPNO, with Dilworth, Secord and Meagher, and later with Freeman Fox, but it proved peculiarly difficult to find a solution in which the engineers had sufficient faith. When the mounting went out to tender late in 1969, the declination bearings were still unspecified. The situation was in fact approaching crisis point, so much so that the possibility of going to an alt-azimuth mounting was seriously raised by Jeffery with individual Board members.[8]

Any such prospect was brought to an abrupt end by Jeffery's untimely death on 2 September 1969. He was a first-class engineer and an excellent leader, and the Project Office was behind him to a man. His loss was a grievous blow, especially because he had been so intimately concerned with the structural aspects of the telescope mounting. It was fortunate that a man with the ability and experience of Harry Minnett was available to take over at short notice. Partly because of Jeffery's death, it became necessary to engage additional consultants to assist prospective contractors and help evaluate their bids. The choice fell on Jeffery's parent firm, Freeman Fox and Partners. They became the consulting engineers for the supervision and general conduct of the drive and mounting contract. One of their staff members, Colin Blackwell, was Consulting Engineer until the telescope was completed; another, Derek Fern, was made Resident Engineer for on-site supervision, and to maintain day-to-day liaison between the manufacturers and the contractor.

Figure 7.5 Project Manager (1969–70) Harry Minnett *(Photo: CSIRO Division of Radio-physics)*

Early in 1970 Blackwell at last was able to tell the Board that satisfactory bearings could be designed, and that there was a London company, Ransome, Hoffmann and Pollard Bearings Limited, which could make them. The bearings, 930 mm diameter, were duly built and installed, with the best possible result, that they have never been heard of again. It is difficult to describe the final bearings in non-technical language. Like other technical subjects which enter this book, bearing design has its own concepts and its own rather impenetrable vocabulary, so we leave it to Minnett:

> In the final design the radial and axial loads are supported by independent bearings. The radial bearing is mounted on an axially-flexible diaphragm and a uniform load distribution is maintained on the axial bearing by means of an annular hydraulic thrust pad. The declination bearings are supported inside the arms of the horseshoe and support a massive steel structure, the heaviest single item of the telescope.[9]

The final optimisation of the polar axis box girders was carried out by Freeman Fox. As Derek Fern put it: *'A computer programme was evolved specifically to solve the many conditions to establish the optimum design. We aimed for a structure as stiff as possible, but which also deforms in such a way that the telescope pointing accuracy varies by the least amount possible as the structure rotates.'*[10]

7.5 Fabrication and erection

The two final competitors for the mounting contract were the British company, Vickers Limited, and the Mitsubishi Electric Corporation of Japan. They provided a striking contrast in styles. Vickers proposed to spend 15 weeks on design and drawing and 52 weeks on fabrication and machining. With Mitsubishi the respective numbers were 44 and 22 weeks, a reflection of the Japanese emphasis on a relatively large proportion of professional engineers. The British naturally wanted to see as much money as possible spent in Britain, and fought hard for Vickers, but Mitsubishi's price was 30 per cent lower, and that proved decisive. Because the Japanese were such important trading partners of the Australians, the Australian Government was understood to be not displeased that so visible and prestigious a contract had gone to their country.

Mitsubishi's real expertise was in satellite communications – large aerial systems, and transistor and drive technology – and it duly won the drive and control contract also, which may have been where its interests really lay. Many of the larger items were subcontracted; in particular, the polar axis structure went to Japan Steel, which also made the polar axis assemblies for the Kitt Peak and Cerro Tololo telescopes. Goodsell has described how on a visit to its Muroran plant he was standing high above the main machine floor watching the massive steel plate, some of it 12 cm thick, being shaped and fitted before welding – heavy engineering with a vengeance. Something discordant caught his eye, and he was astonished to realise that half the workforce were women: gone for ever his vision of dainty Japanese ladies inserting one last branch into an exquisite ikebana arrangement.

The mounting contract was let in November 1970, with the polar axis, base frame and gearboxes as its main components. The mounting was required to be completely shop assembled and tested under full load before leaving Japan. The shop tests began in September 1972 with the assembly and alignment of the base frame. The polar axis – horseshoe, coudé struts and north bearing – was put together and lowered on to the base frame, followed by the declination axis centre section. A concrete counterweight of 64 tonnes simulated the missing tube, mirror and cell, which were still being completed in the Grubb Parsons works in Newcastle–upon–Tyne. The polar axis was carefully balanced, and when it was finally floated on its hydrostatic pads the 280-tonne mass could be easily rotated with one hand. This was in January 1973. The whole was then dismantled, packed and dispatched to Australia, in a ship aptly named 'Cosmos'. Final site assembly began in April, to be completed in August 1973, with the mounting then ready to receive the tube.

The final assembly was fascinating. The telescope of course was shipped out in pieces, their dimensions matching the capacity of the dome crane (46 tonnes) and the size of the shaft (5.5 metres square) up which they were lifted to the main telescope floor. The horseshoe for instance was divided into a middle section and two side

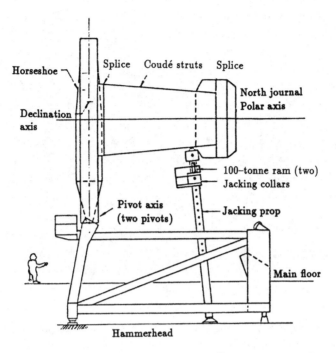

Figure 7.6   How the polar axis was assembled on site (from a diagram by Derek Fern)

pieces, as indicated on Figure 7.2(b). The assembly of a large equatorial telescope is complicated by the requirement that the polar axis must point at the celestial pole, and must therefore be inclined to the horizontal by an angle equal to its own latitude, in this case 31°16′. This makes the process of fitting the polar axis to the bearing pads a very delicate one. With the AAT these problems were bypassed by assembling it with the polar axis horizontal, beginning with the horseshoe (see Fig. 7.6). Two brackets were first attached to the base frame, between the oilpads, and the centre horseshoe section firmly anchored to them, accurately in the vertical. One arm was then lowered on to it, the splices were positioned with internal jacking screws, and clamped in place by high-tension bolts; no dowel pins were used. The process was repeated with the other arm. The 'lip' between the two sections of the bearing surface was no more than 40 microns, and could barely be felt when a finger was run over it. Later it was stoned off. This philosophy, which was carried right through, was to make all final adjustments on site, the components having been previously tested as thoroughly as possible in the shops where they were made. One consequence is that the sections of the polar axis are held together by friction, common practice in structural but not in mechanical engineering.

Following the assembly of the horseshoe, the box girders were bolted together on the floor, then added as one piece, and the final stage was to bolt on the north journal. In

Figure 7.7   Assembling the AAT: the third section is lowered into position to complete the horseshoe *(Photo: Don Collins)*

this way, piece by piece, the polar axis went together. When complete it was lowered into its correct position. The hydrostatic bearings were then pushed into light contact with the runners and shims inserted, thus making the bearing configuration fit the polar axis and not vice versa.[10] The bearing surface of the horseshoe is a cylinder, and the north journal is a part-sphere. During a trial assembly it was shown that the centre of the sphere lay within 100 microns of the axis of the cylinder. And when the the declination axis was finally assembled, it was found to be orthogonal to the polar axis to within 1.5 arc seconds. These figures are typical of the precision achieved throughout the whole manufacture.

The gearboxes and main gears had been installed earlier, and the delicate operation of meshing them was carried out in January 1974. The gearing had already been put through a rigorous test schedule, which began in Japan in February 1973, and was carried out before a bevy of experts which included the Project Office's consultant, W. Finn, from Marconi Radar Systems, and representatives from the Project Office, Maag Gear-wheel and Mitsubishi. Single flank tests revealed some tip interference between the pinions and the main gearwheels. It was alarming at the time – 'sparks everywhere' as someone said; the trouble was relieved by stoning a few microns off the tips of the gearwheel teeth. More problems were in store, but fortunately none proved irremediable.

The wiring was an immense job. There are almost 200 individual motors on the

Figure 7.8   Assembling the AAT: working on the tube centre section after it had been secured in the polar axis. Note the massive struts which are the distinctive feature of the AAT polar axis *(Photo: Don Collins)*

telescope, many encoders ('if it moves, encode it') and innumerable switches, interlocks, status indicators and so on. All circuits had to be completed and checked before the telescope could move; this did not happen until early 1974. Mitsubishi had estimated that it would have taken two weeks in Japan and four weeks on site. Actually, using Australian labour, it took four months. The control consoles and cubicles had to be incorporated into the system, and the first interfacing made with the computer. On-site acceptance tests commenced in mid-March, and the certificate for practical completion of the mounting and drive was granted 23 April 1974.

7.6 The control computer
The subject of drive and control divides neatly into two. The first part, the drive, concerns the gearboxes, motors, servo loops, encoders, power supplies, cables and switchgear; much of it has already been described, if at times cursorily. The second, control, includes the computers, the control console and a great deal of software. The drive is the province of the electrical engineer, control the province of the systems engineer and the programmer. It will be appreciated that the average astronomer takes no more interest in the engineering aspects of his telescope than the average driver in the engineering aspects of his car. What does concern him is that the image be sharp, the drive smooth, the pointing accurate, and that the telescope be convenient to use and amenable to his needs: in short, user-friendly. If the drive is none of his business, control most certainly is, and his close participation in all matters relating to it was essential.

We have described how following the urgent recommendation of the Project Office, the JPC decided to incorporate a computer as an integral part of the control system early in 1968. The computer would control the movements of the telescope, providing corrections for refraction, flexure, gearing errors, and the inevitable telescope maladjustments. It would also carry out mean-to-apparent place and similar calculations, control the dome and windscreen, automatic guiding, digital displays, and possibly carry out a certain amount of instrument control and on-line data-processing.

At that time much of this was unknown territory. It takes an effort to realise how primitive computers then were, and how little experience had been gained in using them for this kind of on-line control. It was necessary to initiate an extensive round of discussions, on the one hand with manufacturers and consultants, among whom the British company, GEC–Elliott Process Automation Limited, played a prominent role, on the other with astronomers. The latter were represented in the first instance by Vincent Reddish (ROE) and Alex Rodgers (MSO). Much of the British input came from the Panel for the Instrumentation of Large Optical Telescopes (PILOT) limited to astronomers younger than 40.[11] Eighteen years later its first chairman, Robert Wilson, became chairman of the AAT Board. And in August 1968 a sizeable conference on instrumentation and computers was held in Cambridge. If its outcome was generally inconclusive it was because planning for more than a few years ahead had been made quite uncertain by the explosive developments, then just beginning, in solid state and computer technology.

Overall responsibility for the computer was given to Maston Beard, who was seconded to the Project Office on a half-time basis from the CSIRO. Beard had just completed the control system for the CSIRO radio heliograph. This instrument was effectively a synthesis array of 96 radio dishes, equally spaced around the circumference of a circle 3 km in diameter. It says something about the technology of the time (the first observation was obtained in September 1967) that the outputs of the 96 dishes were combined not with a computer, as would have been taken for granted a few years later, but with a hard-wired system. Under Beard, Tom Wallace, a systems engineer from the Royal Observatory, Edinburgh, and Graham Bothwell, a young Australian, were made responsible for the software and hardware respectively.

Discussions continued until early 1973, when a list of tasks was drawn up which formed the basis of a design study with GEC–Elliott. The aim was to find, in detail, how these tasks were to be implemented, and what was needed in the way of hardware; its principal outcome was a specification for the computer. It had become clear that for simplicity, and to accommodate future instruments with unpredictable requirements, it would be preferable to have two computers, one for the telescope, the other for instruments. Two Interdata Model 70s were therefore ordered, chosen for their speed and flexibility, their proven real-time operating systems, and because good local support was available. The first was delivered late in 1972. It became the telescope computer, and programming with it began at once, though it was difficult to make real progress until the telescope was working, and the computers and programmers were installed at Siding Spring. This happened in January 1974.

Computer control of this sort differs from data-processing or number-crunching in two main respects. First, it has to manage a continual stream of data, inwards from the encoders, the clock, the autoguider and the console keyboard, outwards to the servo drives, switches and display devices like console videos; it must interface simultaneously with a large number of peripherals. Second, it must handle many tasks concurrently, all of which are, in effect, competing for computer time. The telescope position and drive processes have the highest priority; they are triggered ten times a second and must be completed within 50 milliseconds. Other programmes fitted in according to their priorities. The procedure is orchestrated and time allocated by the so-called real-time operating system (RTOS). One of the principal reasons for choosing the Interdata computer was the known quality of its RTOS, and the fact that it could be expanded and adapted to suit the particular needs of the AAT. The Interdata RTOS was in fact considerably expanded, and is quite central to the whole operation.

Because the control system is probably the best-known feature of the AAT and is central to the efficient running of the telescope, we describe here in simple terms how the telescope tracks on an object, and how it is made to slew rapidly to the next object. A more technical account has been given by Straede and Wallace.[12]

Tracking is effected through the incremental encoder, which it will be recalled has a bit size or resolution of 0.05 arc seconds. Consider the movement in right ascension. For a perfect and perfectly adjusted telescope working in a refraction-free environment, the tracking rate is 15 arc seconds per second of time. The tracking program works in steps of exactly 0.1 time-seconds duration. For this perfect telescope, at every step the program will cause the rate generator to feed 30 pulses into the servo system, whereupon the telescope will advance 30 steps of 0.05 arc seconds, or 1.5 arc seconds in right ascension. But the telescope is not perfect. Both the polar axis and tube can be expected to bend, by amounts which will depend on the hour angle and declination; there will be maladjustments – the polar axis, for instance, is unlikely to be pointing exactly at the pole; and because of refraction in the earth's atmosphere, astronomical objects will appear closer to the zenith than they really are, by amounts which increase with zenith distance. All these effects are taken care of by the computer, which calculates both the current position and that 0.1 second ahead, then injects the difference between the two into the rate generator. Any fractions remaining after the integral number of pulses has been fed to the servo system are carried forward to the next tenth of a second.

Note that rates as such are not computed: the system works solely on first differences, or more exactly, on the integral numbers of encoder bits in the first differences. The actual process is more elaborate, but it produces an extremely smooth drive, avoids the possibility of certain subtle numerical errors, and lends itself naturally to operations like beam-switching and the execution of spiral and raster scan patterns. Tracking in declination is controlled similarly, except that there is no underlying 15 arc seconds per second rate. While the computer is superintending the tracking, it has also to ensure that the dome and windscreen are correctly placed with respect to the telescope. This means calculating their positions about once every five seconds, and feeding the resulting signals into their control servo systems. The display on the console video has also to be updated once a second, to 0.1 seconds in right ascension and one arc second in declination. Finally, if the autoguider is working, it places an additional heavy load on the computer.

When in the slew mode the computer works through the absolute encoders and, except for the final phase, the control servo loop is bypassed completely. Given the position of the next object, its distance is found and speeds assigned so that the slew times taken along each of the axes will be identical. The axis with the longer distance is traversed at the maximum speed of 45° per minute, the other axis at the appropriate fraction of that speed. The speed is controlled by a voltage, generated via the digital-to-analogue converter; near the target it decreases as the square root of the target distance. The telescope decelerates at the last possible moment, and when within about 100 arc seconds of the star it switches into a setting mode which uses the normal control servo system.

The software which achieved all this, and more, took five programmer-years to write, and was packed into 64 kilobytes of core store and two 2.5 megabyte discs. Plenty of desk-top computers are now available with more storage than that.

The scheme works extremely well, and on the first trial the sceptics, who were numerous, were converted immediately. As Straede and Wallace describe it:

> A button is pressed to initiate the slew. The telescope immediately accelerates and in a few seconds is moving at full speed towards the target. At the last possible moment the telescope slows down rapidly but smoothly. On the TV monitor – showing the central 70×120 arc seconds of the f/15 Cassegrain field – the star abruptly appears, pauses momentarily, and darts to the centre of the screen, settling in a second or two. The distance of the star image from the axis marker is now typically 3 arc seconds. It is inconceivable that a human operator could match this performance.[12]

The control system, in which the computer plays such a essential part, is almost certainly the best-known feature of the AAT.

7.7 The Astronomers Working Party
Following a proposal submitted to the March 1971 meeting of the JPC, astronomer

participation was formalised by setting up an Astronomers Working Party, often referred to as the Rodgers Committee. Its membership was Rodgers (Chairman), Beard, Gascoigne, Reddish and Redman, with Project Office engineers as needed. The proposal embodied a good deal of the thinking of the time, and we quote from it:

> The acquisition and guiding system and the computer are the parts of the telescope which are critical to the astronomer, directly affecting his observing efficiency, and bearing intimately on almost every operation he performs. . . Now that work has begun in earnest in this field, closer relations and more efficient communications are clearly needed between the astronomers and the engineers. . .
>
> Underlying these proposals is the realisation that we will be competing directly with the ESO and Cerro Tololo telescopes, instruments in good climates and backed by large, powerful organisations. Our principal advantage is the technological superiority of Australia over Chile, and the ability this gives us to operate at a more sophisticated level. To capitalise on this advantage, we need a high degree of development of the control, guidance and acquisition system. This can be done only with the intimate and detailed concern of the astronomers, who must be prepared in addition to carry a substantial part of the overall responsibility.[13]

The telescope did indeed capitalise on the technological advantages of Australia, sometimes in unforeseen ways, and overall quite as effectively as could ever have been hoped.

The Rodgers Committee cast a wide net. At different times it considered the prime focus camera, the Cassegrain instrument head, the control console, the autoguider, automatic focussing, television cameras for field viewing, and broader aspects of the instrument programme. A number of these topics are discussed here; others are deferred until Chapter 10, on commissioning. The Committee worked in various modes. For instance, it had full-scale wooden mock-ups of the prime focus and Cassegrain cages built and set up in an annex to the Mount Stromlo workshop. They could be tilted through various angles and to different azimuths to simulate actual working conditions. Margaret Burbidge, a recognised expert in such matters, was invited to sample the prime focus cage. Her verdict was awaited with something of a breathless hush, but all was well. Both of these mock-ups, and the reactions to them, were a considerable help to the engineers. A similar full-scale mock-up was made of the control console (Fig. 7.9). On a more technical level, the first trials of the autoguider were made on Mount Stromlo telescopes, and prototypes for the support pistons for the primary mirror were built and tested in the MSO workshops.

The need for autoguiding had been emphasised early, in the Ford–Minnett report, and it was decided in 1970 that the instrument heads at both prime and Cassegrain

Figure 7.9   Robert Dean, computer technician, at the control console of the AAT

foci were to be equipped with autoguiders. Guiding is a time-honoured astronomical procedure, made necessary by the inability of a telescope, no matter how well made, to follow the motion of a star to within the requisite few tenths of an arc second, especially over a long period. A star near the object under observation is selected and centred on the crosswires of a guiding eyepiece; the observer keeps it there by applying delicate corrections to the telescope with control buttons on a handset. Alternatively the star is centred on a spectrograph slit, which may be only an arc second or less wide, and kept there in the same way. Manual guiding requires concentration, patience, and the ability to keep awake and to endure the cold. It is demanding, extremely boring and not at all efficient.

However desirable it may have seemed to several generations of astronomers, autoguiding was an impossible dream until technical developments in the mid-1960s, including the need for star trackers in early satellites, suddenly made it practicable. By 1972 the Rodgers Committee found that there were several options between which it had to choose for the AAT's first autoguider. It decided on a KPNO design, a so-called image scanner, based on an image dissector phototube. The final instrument is described by Kobler and Wallace.[14] Here we discuss only a few salient points.

The guider has two modes of operation. Once the star has been placed manually in the guider field, the guider goes into an initial or acquisition mode in which it locates the star within a raster scan which covers a square of 24 arc seconds per side. After location, the autoguider automatically enters a guiding mode, which uses a cross-shaped scan with arms 12 arc seconds long, repeated ten times a second. A control loop continually centres the scan on the star, and any departure of the scan from its nominal position creates an error signal. Suitably scaled, this is fed into the drive via the computer to eliminate the guiding error. The process is displayed on a video screen in the control console, showing the position of the star relative to the screen centre. The scanner makes heavy demands on the control computer – it was initially proposed that it have its own computer – but it has worked extremely well. There is no doubt that the image quality of plates taken with the autoguider is significantly better than those taken without it, especially in good seeing. Besides its guiding function, the system has been a useful monitor of the mechanical performance of the telescope, since it gives an immediate impression of the response of the servo system.

The fainter the objects which can be observed, the better pleased will be the astronomers. Especially at that time, astronomers set great store by their ability to push telescopes to the faintest possible limits. But observing faint stars, particularly those too faint to see by eye, is difficult and time-consuming, with often as much time spent in the nerve-wracking process of finding the star as in actually observing it. If the astronomers were united on one point, it was that with the AAT the acquisition of faint stars would be made as efficient as possible.

A common way of going about this was called offsetting. The object to be observed is located on a photographic plate, a suitable reference star, bright enough to be identified with certainty, is selected nearby, and the distances from the bright star to the faint object are measured in the N-S and E-W directions to a fraction of an arc second. These measurements are the offsets. The reference star is then centred on the slit, and the telescope is moved by the calculated offsets, but in the opposite sense, whereupon the faint object should have replaced the star. The reference star is then sometimes used as the guide star. The procedure works, but needs care both in setting up and in execution, and presupposes the existence of a deep enough photographic plate. However it does not require great precision in the absolute pointing accuracy of the telescope.

It was clear that at least in the first years the principal observing station would be the Cassegrain focus, and that the Cassegrain observing head therefore would be a critical interface between the observers and the telescope. Accordingly, a great deal of attention was paid to the design of this head, particularly to the acquisition of faint stars by offsetting. Thus the head was provided with two remotely controlled probes, a probe being a small periscope which redirects the light of the reference or guide star to the eyepiece or the autoguider. The probes could be driven in two dimensions to cover the entire field, about half a degree square, and their positions were encoded, with control and readouts centred on the console. The Cassegrain head had to be able to carry instrumental loads approaching a tonne, often with a very asymmetrical weight distribution. As at the same time the guide probes had to work to high accuracy, very little flexure could be tolerated, and the design of the head was a major undertaking. The design was carried out by engineers in the Project Office, while the instrument was built by the SIRA Institute, at Bromley, near London.

However, the complication and extreme precision built into this observing head turned out to be unnecessary, for reasons which could hardly have been foreseen. Nobody realised just how accurately the AAT would point and guide, still less the extent to which this high accuracy would affect traditional observing procedures. At about the same time, television as applied to the detection of faint objects made great strides, basically by way of methods for integrating single frames over times of several seconds, and suddenly one could detect, on a video screen, much fainter objects than was previously possible. Here the AAT was helped substantially by the experience gained at Mount Stromlo Observatory. In addition, a scheme had been built into the control system which allowed highly accurate differential movements of the whole telescope to be made from the control room console. It was based on the high resolution, 20 bits per arc second, of the incremental encoders. During observing, one keyed in the required offsets, in the N-S and E-W directions and to the nearest tenth of an arc second, and pressed the button. The telescope moved quickly and unerringly to the required position, and observations of the unseen object could begin at once. Though intended initially for offsets and beam-switching, the procedure was quick, convenient and accurate, and soon made its way into the computer programs.[15]

The prime focus head, or camera, was conceived by the Committee as a basic instrument of the traditional type, intended largely for direct photography, and to be used until a demand for something more sophisticated arose. Completely manual, it was designed in the Project Office by Peter Gillingham, his assistant engineer Peter Hewitt, and Ben Gascoigne. These cameras are used in total darkness, and the observer has to proceed mostly by sense of touch. Hewitt was taken observing on the 74-inch and 40-inch telescopes at Mount Stromlo Observatory, and on each instrument went through the routine of focussing, loading and changing plates, and guiding. As good engineers do in such circumstances, he learnt quickly, and the resulting AAT camera, a good sturdy instrument, seems likely to remain in service indefinitely.

By the end of 1973, commissioning of the AAT was imminent and the Rodgers Committee ceased functioning. No new proposals could be considered until the telescope was in regular operation, and by then it would be under a different management.

Notes to Chapter 7

1 cf. Hoyle F. (1982). *The Anglo-Australian Telescope*, University College Cardiff Press, pp.6 & 26. This account differs somewhat from Hoyle's. The AAT Technical Committee, especially Redman and Gascoigne, had little to do with recommending the KPNO design.

2 Letter from Redman to Burbidge (RGO 37/160) 1972. 'I think it is a mistake to assume too readily that US telescope design is all sweetness and light, and British design the opposite. To the best of my knowledge every postwar US telescope has had its difficulties, optical or mechanical, some of them serious, and not all due to plain bad luck.'

3 Letter from Bok to Woolley (RGO 47/2)

4 In 1968, as a result of differences among the Canadian astronomers as to whether the telescope was to be located in Western Canada or Chile, the entire project was cancelled. Several years later it reappeared as the Canada–France–Hawaii Telescope, to be located on Mauna Kea and administered somewhat on the lines of the AAT. See G.J. Odgers and K.O. Wright (1968). *Journal RAS Canada* **62**, 392.

5 Minnett H.C. (1962). *Sky and Telescope* **24**, 184; Bowen E.G. and Minnett H.C. (February 1963). *Proc. IRE (Aust)* **24**, 98

6 Trumbo D. (1966). *The Construction of Large Telescopes* ed. D.L. Crawford, Academic Press, p.131. An early use of a digital servo system was on the Mark II radio telescope at Jodrell Bank.

7 Bertin B. (1971). ESO/Cern Conference on Large Telescope Design, ed. R.M. West, p.395

8 Hoyle F (1982). *Anglo-Australian Telescope*, University College Cardiff Press, p.12

9 Minnett H.C (1971). *Proc. Astron. Soc. Aust.* **2**, 2

10 Fern D. (Dec 1975). *Metal Construction*, pp.600–602

11 PILOT was set up by the APGC in 1968. The original membership was: R. Wilson (Chairman), A. Boksenberg, R.F. Griffin, P.W. Hill, B.E.J. Pagel, A.D. Petford, V.C. Reddish, D.B. Shenton, P. Strittmatter. M. Feast of Pretoria was a corresponding member, and J. Ring attended several meetings by invitation. See *Observatory* (1971) **91**, 89

12 Straede J.O. & Wallace P.T. (1976). *Publ. Astron. Soc. Pac.* **88**, 792

13 Gascoigne S.C.B. *Proposal for an Acquisition, Guidance and Computer Working Party*, tabled at the first AATB meeting.

14 Kobler H. & Wallace P.T. (1976). *Publ. Astron. Soc. Pac.* **88**, 80

15 This was a good example of an astronomical contribution to the design. The engineers took quite some convincing, mostly by Gascoigne, that there would be sufficient demand for this type of offsetting to make it worth installing.

8 Telescope or observatory?

8.1 Background to the AAT management

At first sight, it may be thought that the problems of management and operation of the Anglo-Australian Telescope would be largely of a technical and logistic nature. It was to be built on a mountain top already occupied by an established observatory and, although it was to be bigger and more complex than any existing telescope in Australia or Britain, experience in the United States had shown that telescopes could be operated successfully in such circumstances. However, the JPC and the AAT Board were to find that deciding how to manage the operation would lead to long, drawn-out and at times acrimonious arguments with the upper echelons of the Australian National University before the matter was resolved. The University's case for operating the telescope was that it had the expertise, it could perform it more economically, and it owned the mountain.

The discussions which took place and decisions made between 1968 and 1973 embroiled several strong and colourful characters and a host of lesser players who embraced and ultimately embodied the experience and traditions of the Australian, British and American astronomical communities into the internal organisation of the AAT. Many of the personalities have left the scene and much of what follows has been reconstructed from the minutes of countless meetings and fleshed out by the recollections of some of the survivors. The minutes, of course, reveal only the substance, rarely the style of the meetings, but the result of this turbulent and difficult period was so important to the success of the AAT that we have covered the period in some detail.

Mount Stromlo Observatory had held a pre-eminent position in Australian optical astronomy since the Second World War. Its principal claim to expertise was that its astronomers had taught themselves to use the 74-inch telescope, and had used it well. Previously, they had extensively modernised their elderly 30-inch reflector, and had virtually rebuilt the old Melbourne 48-inch telescope. In addition, they had constructed, among other instruments, a highly successful coudé spectrograph. In these respects they had a clear lead on any other British group. On economic grounds Mount Stromlo argued strongly that it was wasteful to build a second observatory to run the AAT, and this had an obvious appeal, especially to the Australian Minister, Malcolm Fraser. However, the Mount Stromlo estimate of $400,000 per year (or 2.5 per cent of the capital cost) for an operation with a staff of 10 or 12 was clearly too low. It no doubt was based on Eggen's experience with the Palomar 200-inch which also had a staff of 10 or 12, but took no account of the much greater sophistication of the AAT, and of physics equipment generally. Even in the 1960s the American standard, as expressed in the influential Whitford Report, was 4 per cent of the capital. Yet nobody challenged Mount Stromlo's figure at the time.

There were other factors which influenced the University's case. Mount Stromlo scientists, especially Bart Bok, had been part of the driving force for a large telescope in Australia. They had been members of various technical advisory committees and of the committee of the Australian Academy of Science which presented the formal submission to the Government in 1965. Scientists and engineers from Mount Stromlo worked with the Project Office during the construction of the telescope, and Mount Stromlo's astronomers formed a large proportion of the likely users of the Australian observing time. Therefore, it was reasonable that the ANU expected to have a considerable say in its running. If the AAT had been a completely Australian undertaking, this no doubt would have happened.

The AAT was in the unique position of being the first large telescope built without any established organisation specified to use and to operate it. In its submission to the Government in March 1965, the Australian Academy of Science put forward a firm proposal for the administrative arrangements: 'A permanent staff consisting of a director, scientific and other personnel is to be employed.' [1] Nevertheless, the Anglo-Australian Telescope Agreement is conspicuously silent on this point, and the absence of any clear provision for management of the telescope left all too much scope for differing interpretations to be placed on the particular clauses of the Agreement. We quote Article 4 of the Agreement which gave rise to so much debate.

1. The Commonwealth Government shall arrange with the University for the use by the Telescope Board of a site for the telescope in the area that is vested in or under the control of the University at Siding Spring Mountain in the state of New South Wales. The terms and conditions of such use shall be as agreed upon between the Telescope Board and the University.

2. So far as practicable and subject to satisfactory arrangements being made between the Telescope Board and the University, use should be made of supporting facilities in existence or to be provided by the University at Siding Spring Mountain and at Mount Stromlo. This does not however preclude the use of supporting facilities elsewhere.

3. The arrangements for the provision by the University of facilities and services for the purposes of construction, operation and maintenance of the telescope shall be such as are agreed upon between the Telescope Board and the University, and the Commonwealth Government shall accord its good offices as appropriate in the negotiations and the putting into effect of these arrangements.

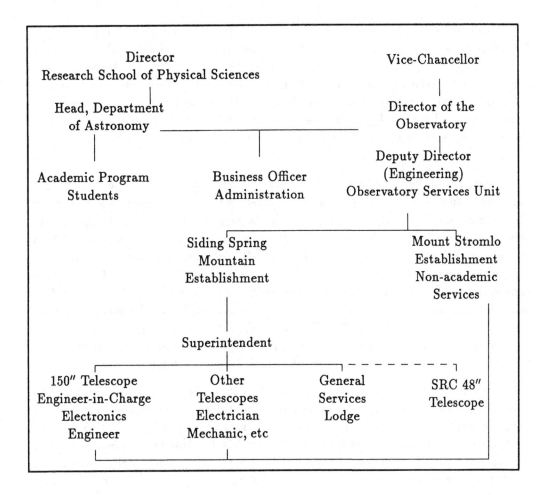

Figure 8.1 Eggen's proposed structure for management of the AAT

The matter began simply enough in 1970. At the fifth Joint Policy Committee meeting in March, Professor Olin Eggen, the Director of Mount Stromlo Observatory and one of the JPC members, presented a humorous paper describing the progress of a fictitious set of applications for telescope time, in an attempt to force early consideration and discussion about the operation and management of the telescope. Some JPC members did not appreciate Eggen's light-hearted treatment of the subject, and K.N. Jones, the Australian Department of Education and Science representative on the JPC, felt obliged to provide a paper which took a more serious approach to the matter. Both papers assumed that the AAT would be a part of the ANU complex at Siding Spring Mountain, which then comprised three telescopes (of 40, 24 and 16 inches) run by a staff of fewer than six people, excluding a small domestic staff which ran the accommodation, known as the Lodge. Allocation of telescope time made by the Board would be passed to the Director of Mount Stromlo and Siding Spring

Observatories, who would make the day-to-day schedules and general arrangements for supporting facilities on the mountain. These papers estimated that the AAT operation would require nine additional staff to be employed by the University and integrated with existing staff at Siding Spring to operate the Observatory and the new AAT as a whole. Eggen's paper, in addition, envisaged an engineer in charge of the AAT who would be the deputy to the Superintendent of Siding Spring Observatory. Eggen's proposed management arrangement is displayed in Figure 8.1.

8.2 The British view

The British members of the JPC had serious reservations. J.F. Hosie, the Science Research Council (SRC) representative, pointed out that it seemed the Director of Mount Stromlo Observatory in effect would be responsible for drawing up the observing schedule for the telescope, and that the proposals did not foresee any staff on the mountain directly employed by the Board. He could not accept this and thought the matter should be discussed at the next meeting. In view of the SRC's experience with the Isaac Newton Telescope, he also argued that some allowance should be made for instrumentation for the AAT when considering financial provisions for future years. Jones argued that the Agreement covered only the running of a telescope and not an observatory and would only admit the possible need for limited provision for instrumentation after the erection phase had been completed.

In anticipation of a request from the Board to the ANU for support facilities, the Vice-Chancellor, Sir John Crawford, had begun considering how the University should organise itself to provide for the new telescope services. In July, R.A. Hohnen, the ANU Secretary, advised Jones that he believed that the University would accept in principle the proposals in Jones' paper to the JPC; Crawford regarded as important the concept of a single observatory with a sharing of facilities and a common staff with the University; he would be willing to discuss these matters with the Board, and would expect to appoint an engineer in charge of the whole Siding Spring Observatory who would be deputy to the Director of Mount Stromlo Observatory. Quite apart from the fact the ANU proposals were not solicited by the JPC, Bowen, its Chairman, regarded them as suggestions which would not appeal to the majority of astronomers in either the British or Australian community.

The sixth JPC meeting in August considered the whole question of management of the telescope in Jones' revised paper which proposed that the ANU would act as the Board's agent, making facilities and services available at Siding Spring and Mount Stromlo. However, it still saw the facilities at Siding Spring and the AAT itself operating as an integrated whole under a deputy director, to be appointed by and in consultation with the Board. The University would provide any additional staff for the new telescope. While these proposals could represent effective and economical arrangements for operating the AAT, the British members of the Committee made it clear that they could not be bound to the proposals without the agreement of the Science Research Council. The interim nature of the JPC, as distinct from the eventual permanent legal body of the AAT Board, made such a course essential and prudent. Moreover, it was known that many of the younger members of the British astronomical community, who would be the principal users of the AAT, were opposed to the subordination of the telescope operations to the Director of Mount Stromlo. Their wish was to have a direct employee of the Board responsible for the AAT.

In September 1970, Hosie explained the situation to the Astronomy Space and Radio (ASR) Board of the SRC. This Board reserved judgment on the merits of the proposals from Jones and requested that the views of young practising British astronomers be sought through the panel on which they were largely represented. This panel, known as PILOT, the Panel for the Instrumentation of Large Telescopes, advised in turn the Astronomy Policy and Grants Committee (APGC), which was chaired at this time by Professor Fred Hoyle, himself a member of the JPC.

During the next few months the essence of the proposals under consideration for the operation of the AAT percolated through the astronomical community in Britain and to the expatriate British community in the USA. There was mounting criticism which varied from general unease that the proposals would give undue power to the Mount Stromlo astronomers, to outright condemnation that the proposals were wrong and unworkable, and that there should be an AAT director responsible to the Board and independent of astronomy at the ANU. Fears began to be expressed, rightly or not, that elements in the University were bidding for a takeover of the operation and control of the AAT. Hosie explained to the critics that it was necessary to distinguish between policy control, which was to be firmly with the Board, and executive control, which under the right circumstances could be entrusted to an agent, but that no decision had been made on the British position. In November 1970 the APGC considered the views of PILOT that the proposals under consideration by the JPC were unacceptable because effective control would be in the hands of the ANU and the status of British astronomers using the AAT would therefore be reduced to that of guests. For equality of status, the AAT would need a director and a full-time staff reporting to the Board. Preferably, the director should be an engineer, not an astronomer, but the APGC on the other hand believed an astronomer rather than an engineer should be in charge. The whole issue was of such importance that it was referred to the ASR Board and the Council. The PILOT report was presented by its Chairman, R. Wilson, to the ASR Board where he found strong support on the main issues, although views were still divided on whether the director should be an astronomer or an engineer. However, support was not unanimous; the Astronomer Royal, Sir Richard Woolley, was alone in believing that the proposals before the JPC would form an acceptable basis for an harmonious relationship with the ANU. Woolley, it should be remembered, was himself once Director of Mount Stromlo Observatory, and as one of its honorary professors retained strong ties with the University.

8.3 The Australian view

It was becoming clear that not everyone in Australia supported the ANU proposals put forward by Crawford and Eggen. As a result of his deep involvement in the original negotiations between the Australian Academy of Science and the Royal Society, Sir Leonard Huxley, the University's immediate past Vice-Chancellor, knew the intentions on both sides and was against the ANU position. Sir Ernest Titterton, Director of the ANU's Research School of Physical Sciences, also was known to be critical of the proposals and wrote a paper[2] on the subject. He supported the idea of a separate director and staff for the AAT, with clearly defined responsibilities for operating and maintaining it, although sharing some ANU facilities at Siding Spring, such as the workshop and the astronomers' Lodge. He clearly disagreed with the interpretation of Article 4 of the Agreement, to which Crawford and Eggen fiercely adhered right to the end. A news article in the *Sydney Morning Herald* on 5 December 1970, reporting an interview with Eggen, described him as '. . . *the man who will be in charge of the new 11 million dollar, 150 inch telescope . . .* ' This did nothing to allay the fears of those who suspected a takeover bid. So far the obvious unease came from Britain. There had been no organised canvassing in Australia of the opinions of practising astronomers outside Mount Stromlo.

Feeling was particularly strong against the Crawford line in the Australian state universities, and institutional rivalry, especially with Mount Stromlo Observatory, was certainly an important factor in the debate. The state universities had nurtured high hopes of obtaining observing time on the new telescope, but saw the possibility of the ANU usurping this time. During this period the two leading figures in the University were Sir John Crawford, the Vice-Chancellor, and H.C. Coombs, the Chancellor, both powerful and successful men, the former in the public service, the latter in academic life. Historically, the state universities remained suspicious of the ANU, which was not only a competitor for funds, but whose members were seen to be well placed to establish useful political links in the national capital. During the 1963 debate of the Australian National University Bill which would allow the University to establish a field station at Siding Spring, Bart Bok's recognition of this useful position is evident from a letter to Sir John Cockcroft, dated 11 April 1963: '. . . *You are probably aware of the fact that the ANU and Mount Stromlo Observatory have many good friends in the House of Representatives, . . .*'

The institutional rivalry was not confined to the state universities. The Mount Stromlo scientists saw their monopoly of optical astronomy threatened by a body which would have a far better telescope than their own, and which would absorb money which they expected would have gone to them. The concept of the astronomical cake being divided into institutional slices was accepted at the time.

8.4 Negotiations with the ANU

In December 1970, Hosie advised Jones that the majority of the British community believed the management of the AAT would best be achieved by appointing an astronomer who should be directly responsible to the Board. While there was no wish or intention to duplicate the staff or facilities on the mountain, specialist staff in computing, electronics and instrumentation would probably be needed for the AAT, and they would report to the manager. As far as possible the AAT and Siding Spring staffs would be complementary.

The members of the Joint Policy Committee met as such for the last time in February 1971 and immediately reconvened as the first meeting of the AAT Board, a transformation made possible by the completion of formalities concerning the Agreement and Australian legislation, described in Chapter 4. Bowen had already advised Jones in January that he doubted whether the SRC would agree to the proposals from Eggen and the ANU, and warned that the matter would need to be taken very slowly. The views of the Australian Board members on the British position towards management and operation varied from neutrality of the Chairman (Bowen), through a wish to compromise (Jones), to outright opposition (Eggen). A compromise suggested by Jones involved the appointment of one Board employee as AAT Manager, who would be then responsible for detailed allocation of telescope time. He would also arrange the management and operation of the AAT by ANU staff and would be under instructions to give the AAT work priority over other tasks. Eggen, however, was totally opposed to any staff on the AAT being direct Board employees.

The first in a long and tedious series of negotiations between the Board and the Chancellery of the ANU took place on 25 February 1971. The University contingent was led by Sir John Crawford, the Vice-Chancellor, with Professor N. Dunbar, Deputy Vice-Chancellor, Sir Ernest Titterton, Director of the Research School of Physical Sciences, and R.A. Hohnen, the Secretary. The main thrust of Crawford's argument for close ANU involvement was based principally on his interpretation of Article 4 of the Agreement, which he believed made the use of ANU support facilities a *sine qua non*, rather than a desirable option. In this he was ably supported by Eggen who was not slow to point out his qualifications for the position and keen desire to run the telescope. While denying that the ANU had any wish to take over the AAT, Crawford enlarged on the many advantages the University could offer non-ANU astronomers in the way of facilities and an academic environment. He did not want to see a separate development on Siding Spring Mountain with what he considered wasteful duplication. If there were an AAT Manager, the ANU would not wish an astronomer to hold the post, since the growth of a separate astronomical staff would then be inevitable. Jones described the compromise he favoured, and Hosie reiterated the British position emphasising that British astronomers wished to use the telescope as joint proprietors, not as guests. Titterton stressed the great importance of the British

having right of access. The meeting ended inconclusively, but with an agreement
that the ANU would prepare a paper for the next Board meeting, containing revised
arrangements and emphasising the benefits of the University's proposals.

Crawford was very pro-Australia. He had lived through a period of Australian
history when British views, especially in foreign policy, had influenced Australian
policy to British advantage. He was first and foremost a politician. In this capacity
his first concern was protecting the interest of the University, which was a completely
Australian interest, and in the same capacity he was not one to brook opposition
lightly. He had held senior political appointments as an economist in Australia and at
the World Bank in Washington, and was highly regarded as an advisor to Australian
Governments of all parties. His style of operation had recognisable hallmarks: though
not unaware of the advantages of a consensus, he liked to have his own way in the
face of opposition, and was accustomed to command the prompt attention of people
in leading positions in the country, particularly government ministers. His manner
was no different when dealing with the problems of the AAT. On 26 March, writing
to the then Minister for Education and Science, David Fairbairn, he advised that
in arrangements to be made between the Board and the University, '. . . *the
ANU must not lose its identity nor its place as a university facility.*' [3] In addition,
he emphasised that it would be wasteful to build up a separate administrative and
service establishment for the AAT.

In the academic world outside the ANU, Crawford was also a politician. An extract
from the minutes of a meeting of the Australian Vice-Chancellors' Committee, held
on 15 June 1971, shows an interesting record, quoted here in full.

22.   Anglo-Australian Telescope at ANU [4]

Sir John Crawford said that there was a misunderstanding in some universities about the Anglo-Australian 150-inch telescope at the ANU's observatory at Siding Spring Mountain, NSW. While it was located within
the ANU, it was not exclusively an ANU facility. It was available for use
by other bodies, particularly universities, and Sir John circulated to Vice-
Chancellors copies of a document relating to the use of the telescope. He
asked that this be brought to the attention of persons within universities,
particularly astronomers, who might wish to have access to the telescope.
Vice-Chancellors undertook to do this.

In July, Eggen produced papers to the Board for discussion with British astronomers.
His interpretation of the ANU position was outlined, the most important points being
that the ANU would have no claim on AAT time not justified by scientific merit, that
full membership of the ANU Observatories would be offered to non-ANU astronomers,
and that the ANU offered the Board on a repayment basis, full staffing for technical,

maintenance and operational facilities. The ANU would not want wasteful duplication of staff and divided conditions of employment and control on the mountain.
Eggen proposed the appointment of a Manager, jointly by the Board and ANU, who
would reside at Siding Spring, but have no access to the AAT for personal observing. The Manager would be responsible within the management structure of Siding
Spring Observatory for the management and operation of the AAT within the dome
and would report on these aspects directly to the Board. On the key issue of control,
it was clear that Eggen intended this to remain in his hands. He suggested that the
AAT and ANU Observatories should be operated and maintained as an integrated
whole under the Director of Mount Stromlo Observatory. Any additional staff for
the AAT at either Siding Spring or Mount Stromlo would be provided by the ANU.
The Board would have the observing schedules prepared on a quarterly basis for the
Manager to implement, acting under the Mount Stromlo Director. Eggen's proposals
clearly did not meet the basic British requirement that the AAT should be under the
Board's direct control.

The second Board meeting in August received a paper from the ANU Chancellery
which proposed arrangements substantially the same as those by Eggen. The Board
also considered this paper, but preferred to have further discussions with the potential users of the telescope in each country. At this point, the Board was seen as evenly
divided on the issue of management: Eggen, Jones and Woolley supported the ANU
proposals; Bowen (Chairman), Hosie and Hoyle wanted an independent staff responsible only to the Board. Woolley, prevented by illness from attending this meeting,
was about to retire as Director of the Royal Greenwich Observatory at the end of
1971 and his successor designate, Professor Margaret Burbidge, was known to hold
views in line with those of her British colleagues.

8.5 Control v. management and operation
Following the August 1971 Board meeting, the views of British astronomers were
canvassed and further developed in discussions in both the APGC and ASR Board.
These discussions enabled Hosie to prepare a paper which could go to the AAT Board
as a basis for negotiation, and which was both endorsed by the ASR Board and had
the support of the great majority of British astronomers. These astronomers regarded it as essential that they should use the AAT as of right, in the same sense as
they had access in principle to the Isaac Newton Telescope at the Royal Greenwich
Observatory, and not come to the telescope as guests. While they appreciated Crawford's offer of co-operation with the two ANU observatories, they could not accept
the basic premise on which the University based its proposals for management of
the AAT. Hosie's paper distinguished between *control* which was the responsibility
of the Board, and *management and operation* which could be arranged through another agency if this were deemed more appropriate. *Control* would go beyond matters
appropriate to Board meetings and would involve day to day decisions requiring the

full-time presence of a direct Board employee, referred to at this time as the Manager. The Manager would be responsible for the control of the operation, maintenance and general running of the AAT. He would be more than a mountain superintendent, and it would be necessary to select the best qualified person available, rather than specifically exclude those who might wish to use the telescope themselves, as did the ANU proposals. It was made clear that telescope staff should be eligible to compete on merit for observing time.

Once again, the fundamental criticism of the ANU Chancellery's proposals arose from the fact that the AAT Manager would be subordinate to the Director of Mount Stromlo and Siding Spring Observatories, the obvious major Australian user of the new telescope. Thus the Mount Stromlo Director would be responsible to the Board for future development proposals for instrumentation and telescope improvements. It was seen as undesirable that these should come through a single major user; rather, they should come from diverse users through the Manager to the Board. The essence of the British proposals was that the Board should determine the number and categories of staff necessary for the effective control of the operations, and those exclusively occupied by AAT activities should be direct Board employees. Others could be Siding Spring Observatory staff working under contract between the Board and the ANU, but giving priority to AAT tasks.

Once Margaret Burbidge, the new Director of the Royal Greenwich Observatory, had replaced Woolley on the Board, she made known her views on the AAT management. No doubt based on her experience as a user of large telescopes in the USA, she suggested:

1. The telescope should be headed by a director who was an astronomer, having a small team of resident astronomers, perhaps drawn equally from Australia and Britain and sharing a fraction of the observing time, say 30 per cent The remaining time would be available for equal sharing between British and Australian astronomers visiting for short periods. It had been shown at the observatories at Kitt Peak and Cerro Tololo that resident astronomers were extremely helpful in assisting short-term visitors and newcomers.

2. The director should be responsible only to the Board.

3. When the AAT was operational and exciting results were coming, collaboration and a merging of interests would occur.

These views, particularly the proposal concerning resident astronomers, were rather more radical than those endorsed by the APGC and ASR Board, although the Chairman of the SRC, Sir Brian Flowers, pointed out they were quite consistent with the way in which large successful international projects in nuclear physics were organised.

The arrival of Margaret Burbidge at the third Board meeting in February 1972 would clearly reinforce opposition to the management proposals advocated by Crawford and Eggen at the ANU.

The management issue had occupied a considerable amount of time of a large number of people. There had been no less than three Board sessions, two meetings with Crawford and ANU representatives, and two informal meetings with the Minister for Education and Science, Malcolm Fraser. At the first meeting with the ANU, Crawford showed no inclination to modify his position. To the surprise of the Board members, he referred to advice received from Alan Sandage, from the Hale Observatories in California, then a visiting astronomer at Mount Stromlo. Sandage, who was recognised as one of the world's leading astronomers, supported the ANU position although having no official status in the affair. At the second meeting, Hosie and Jones produced a statement describing what the Board saw as an alternative to the ANU's proposal. The majority of the Board members now believed the AAT would be of such a size and sophistication as to overshadow all other facilities at Siding Spring. The case for a full-time director was even stronger. This director should be the best available astronomer, and should have as direct Board employees such supporting staff as necessary for the AAT alone. Other supporting staff, if appropriate, could be shared with the ANU. Clearly, this statement was not welcome by Crawford and the ANU.

It was equally clear from the Board's first meeting with Fraser that his sympathies on the management issue lay with the ANU. The Minister shared the rigid and narrow interpretation of Article 4 of the Agreement, which went close to implying that the ANU held an interest as if it were a principal party to the Agreement. He impressed upon the British Board members that the Australian Government had spent a large amount of money supporting astronomy at the ANU, and that it did not intend to spend yet more by permitting the Board to establish its own independent operating facilities. Later, at his office in Parliament House, Fraser reiterated that the Australian Government would not duplicate its expenditure on optical astronomy. This ministerial departure from neutrality while the matter was still under discussion alerted at least the British Board members to the political difficulties which might lie ahead, but did nothing to deflect them from the course they were convinced was best for astronomers of both countries.

On 29 February 1972, Sandage wrote Crawford an extremely lucid letter explaining why there must be either total integration of the AAT operation with Mount Stromlo and Siding Spring Observatories or divorce. He wrote '. . . *Questions of pride and national interest have their solutions only if scientific success is achieved.*'

After the meetings with the Board in Canberra in February, Hohnen, the Secretary of the ANU, wrote to Jones that he regarded Hosie and the British astronomical establishment as '. . . *looking to Siding Spring Observatory as a base for a British operation of no small magnitude'.* The University was starting to fear the establishment of a predominantly British Southern Observatory at Siding Spring which would include a 60-inch telescope and a millimetre wave dish.

Aware of the possibility of a dangerous clash of opinions, exchanges became more frequent between the heads of the two Government agencies, Sir Brian Flowers, SRC Chairman, and Sir Hugh Ennor, Secretary of the Australian Department of Education and Science. Flowers had visited Canberra in November 1970 and at that time was already expressing views about suitable people to fill the position of AAT Director. In addition, he told Hohnen that he held Eggen in high regard as an observational astronomer, and the possibility of Eggen as AAT Director could be entertained, but not in his capacity as Director of Mount Stromlo Observatory. However, an article in the *Sunday Australian* on 26 March 1972 which implied that the AAT was being constructed especially for the benefit of Eggen and the ANU must have cast doubt on Eggen's suitability for the directorship.

8.6 The La Jolla meeting

Still there was no unanimity within the Board on the question of management, and a special meeting was held on 27–29 April 1972 finally to resolve the issue. The progress in construction of the telescope itself was bringing home to everyone concerned the need for speed in settling the arrangements for staffing, commissioning and operational control of the AAT. Jones suggested the desirability of having this meeting on neutral ground and, as a convenient half-way point, selected La Jolla in California. This city was to give its name to a major Board decision. Once again Hosie presented a paper, on this occasion amplifying the views of the majority of the Board members on the issue before them. He pointed out that there would be three different bodies administering facilities on Siding Spring Mountain, the bi-national Board of the AAT, the ANU with its own telescopes, and the SRC responsible for its Schmidt telescope to be built under a direct agreement with the University independently of the AAT. A single unified, multinational body was not feasible because neither the ANU nor SRC (with its investments of £3.5 million in the AAT and £1 million in the Schmidt telescope) was prepared to lose its identity on the mountain. The SRC regarded the AAT in the same light as its other internationally financed scientific projects, which were, and were seen to be, internationally controlled. Full co-operation with national and private bodies working alongside such projects had neither disturbed the principle nor presented problems. The ANU proposals, on the other hand, continued to conflict directly with this principle and would give control of the AAT to the Director of Mount Stromlo Observatory.

At La Jolla, Eggen pressed the ANU Chancellery line, arguing that the Agreement provided only for *a telescope* and not *an observatory*. He relied on Article 2 of the Agreement, the relevant part of which reads:

1. The Contracting Parties shall cause to be manufactured, constructed, operated and maintained, by the Telescope Board, an optical telescope and associated facilities and services.

The British regarded Article 2 of the Agreement as capable of being interpreted to provide an observatory, and referred to the Radcliffe Observatory in Pretoria as an example of an observatory with one telescope.

The proposals supported by the majority of the Board and refined after much discussion were summarised in Hosie's paper. A director, who would be the best available astronomer, would be appointed and employed by the Board. There would be six support astronomers, their telescope time allocations coming evenly from the equal allocations to each partner. Technical tasks required for the AAT alone would be carried out by technical staff employed directly by the Board. The balanced group of seven astronomers, each resident for a sufficient time to become thoroughly familiar with the telescope, would provide invaluable support for visiting, short-term users. This arrangement would not involve additional expense, since it was envisaged, even in the ANU proposals, that there would be a permanent AAT representative on the site, and the six astronomers would come naturally from the two communities.

The proposals in the ANU Chancellery paper prepared for the La Jolla meeting did not advance towards the Board position and, indeed, were if anything less accommodating. Eggen produced a personal paper which proposed in effect a complete takeover of the AAT, the total operation on Siding Spring being staffed by key people from Mount Stromlo. In anticipation of this proposal being accepted, he had written to Hohnen on 10 April to set in motion arrangements for Mount Stromlo's chief engineer, Ruting, to spend two months in the USA (at Hale, Kitt Peak and Lick) '. . . in connection with Mount Stromlo's preparation for operation of the 150-inch reflector'.

Bowen, the Chairman of the Board, reminded members that the original submission in favour of the AAT project from the Australian Academy of Science to the Government envisaged a permanent staff of a director with a scientific and other supporting personnel. He also reported on his recent discussions with four very senior and experienced scientists in the USA: H.W. Babcock, J.L. Greenstein, I.S. Bowen and L. du Bridge. All of these favoured an independent director and staff for the AAT. Du Bridge, past President of the California Institute of Technology, advocated strongly the need to avoid the problem of a director being answerable to two masters. This had been the case with the Director of Mount Wilson-Palomar, who was in effect answerable to the Carnegie Institution of Washington and to Caltech.

The discussions at the La Jolla meeting were by all accounts at times very heated and acrimonious. Indeed, Hosie reported to the SRC that it was the roughest meeting in which it had ever been his misfortune to be involved. Bowen described it as the most difficult, hard-fought meeting he had ever attended, let alone of which he had been Chairman. This was only Margaret Burbidge's second meeting, and Bowen described how, in the course of bitter argument, she performed superbly. Not only was her view on running large astronomical instruments delivered with clarity, but she helped maintain the decorum of the meeting which, in her absence, could have got out of hand. Nevertheless, it proved impossible to achieve a unanimous decision about the AAT management arrangements, and because of the pressing need to move forward, a vote was taken. Bowen, who had not broadcast publicly his views until now, long believed that it was customary for all big telescopes to have a director. A man who was an outstanding professional and responsible for overall operations was essential. Two Board members, Eggen and Jones, opposed a resolution which favoured an independent director. The resolution was carried by four votes to two. We quote it here in full in view of its importance to the whole history of the AAT.

> In order to discharge its responsibilities for the control, operation and maintenance of the telescope, the Board considers that it should appoint a Scientific Director and appropriate support staff, both scientific and technical, employed by and answerable only to the Board.

This was the only event which has ever prompted the Board to take a vote. In all other matters, both before and afterwards, decisions have been reached by consensus.

Once the decision was conveyed to the ANU, Crawford wrote to Ennor, Secretary of the Department of Education and Science on 29 May 1972, stating that the University had gained the impression that the Board intended to establish a fully staffed observatory to support the AAT. He claimed this was inconsistent with the understandings on which the ANU agreed to make the Siding Spring site available, and that the Board's decision was not authorised by the terms of the proposal put before, and approved by, the Australian Cabinet in 1967. He further tried to argue that the attitude of Hosie, the SRC representative, was inconsistent with the terms of a detailed memorandum between the SRC and University in respect of the UK Schmidt Telescope to be built at Siding Spring. He wanted to see the Minister.

8.7 The Ennor meeting
Ennor advised Jones that he suspected Crawford's letter had a bias. Nonetheless, this reaction from the ANU may have impressed on Ennor that it was indeed time to consult the whole Australian community of astronomers to ascertain *their* views. The meeting of leading Australian astronomers which convened in Canberra on 11 July included scientists from Mount Stromlo, CSIRO and the state universities and observatories. Many who attended were open-minded on the issue of management,

as the ANU had a good reputation in operating big telescopes. But when Eggen addressed the gathering, his intransigence was evident to any management proposal which failed to give his University a pre-eminent position. This marked the turning point. The scientific heavy-weights were against the ANU. Although no vote was taken at the meeting, Ennor invited those present to write to him with their views. Of the 17 present, of whom three were Fellows of the Royal Society of London, 14 favoured an AAT with an independent director and staff. None of the eight Fellows of the Australian Academy of Science supported the ANU. The majority opinion can best be summarised from the correspondence of B.Y. Mills, Professor of Physics at the University of Sydney, to Ennor on 11 July 1972. Mills wrote that while he went to the meeting feeling there was something to be said for both sides, he '. . . *came away convinced that an independent director of high scientific standing was absolutely essential'.*

The overwhelming opinion of the Australian astronomers had little effect on Crawford. On 14 August, he led the ANU representatives at a meeting with Malcolm Fraser and Ennor. Hohnen argued that support for astronomy in the ANU had been represented to Cabinet in 1965 as the basic case on which the Government decided to proceed with the 150-inch telescope project; further, that the Australian Government had never intended to establish a separate observatory. Ennor, on the other hand, was quick to identify this line of argument as quite erroneous and harmful to future relations between the ANU and Australian astronomers. Directly after this meeting, Crawford wrote to Ennor, '. . . *To appoint a first class staff* [to the AAT] *would almost certainly denude ANU staff and throw in doubt the ANU's ability to maintain a strong Department of Astronomy.'* As it happened, the opposite was true. Later, when the Board was recruiting its own staff, it tried hard to attract people from the University but without success, mainly because it could not offer tenured positions, although the inability to offer tenure was not clear in the early period. The closest anyone from Mount Stromlo came to staying with the AAT was Ben Gascoigne, the astronomical advisor to the Board who was seconded in 1974 for 18 months as Commissioning Astronomer.

8.8 The Thatcher–Fraser meeting

A visit to Australia by the British Secretary of State for Education and Science, the Right Honourable Margaret Thatcher, was planned for August 1972 when she would meet her opposite number, Malcolm Fraser. Clearly there was the possibility that they might discuss the AAT management issue. In fact, it was the first item on Thatcher's agenda in Canberra. The meeting convened in Fraser's office on 30 August. Also present were Bowen, the Board Chairman, Ennor and Jones. With preliminaries completed, the discussion concentrated on the question of the AAT management. As exchanges progressed, it was evident Margaret Thatcher had been briefed fully, had done her homework well, and was completely aware of both the British and Australian views. Fraser emphasised that the Australian Government always intended the AAT to be integrated with Mount Stromlo and Siding Spring Observatories for operational support. He claimed that the Board's decision at La Jolla was not conclusive, that the Australian Government was not prepared to provide funds for supporting facilities. It was time to prevail upon the Board and the University. Thatcher recognised the Board's independence and was reluctant to interfere. The powers under Article 6 of the Agreement were given to the Board to exercise, and consultation with the ANU did not give the University any right to control the Board's decisions. She regarded any ministerial intervention as a drastic step and tantamount to a vote of no confidence in the Board, which in the British opinion had acted within its charter. Her arguments were succinct and comprehensive. She was forthright in her recommendations for total control of the AAT by the Board and completely against a situation where this passed to another body. At the conclusion of the meeting, Bowen recalled a previous meeting of vehement exchanges with Fraser after the La Jolla decision. He had tried then to explain that the Board had been appointed not by one, but by *two* Governments with authority to decide the matter of management, and that it would do this with due regard to all the issues.

The result of the meeting between Fraser and Thatcher was a joint statement to the Board and the ANU, exhorting both to reach agreement within the terms of the inter-governmental Agreement, so that the Board could carry out its responsibilities for the control and operation of the telescope. Discussions planned for the following October were regarded as critical, and it was suggested that a joint paper from the Board Chairman and the Vice-Chancellor, identifying points of disagreement would be helpful.

While the Board was handling problems in Canberra, the AAT came under attack from another quarter. The 8 September issue of *Nature* contained a letter by Geoffrey Burbidge[5] in which he attacked many aspects of British scientific policy. He claimed, among other things, that the decision of the British Government to build jointly with Australia a large telescope on Siding Spring Mountain had been made without proper consideration of observing conditions in Australia or elsewhere. For a short time the pressure was reduced on the management question as the Australian community, and particularly Titterton, argued in defence of the ANU site testing programme[6] carried out in the 1960s which resulted in selection of Siding Spring as the site for the University's telescopes and later the AAT.

8.9 Search for a Director

The fourth Board meeting took place in Canberra in October 1972, when the Board held three meetings with the ANU Chancellery. After careful consideration, it re-affirmed its decision taken at La Jolla, which was supported now by the overwhelming majority of British and Australian astronomers to appoint a director and supporting staff. The Board expanded its views in a paper presented to the ANU, with the hope of persuading Crawford to accept the decision and to co-operate closely in making the maximum use of University facilities and services. It was intended that the director should be the best available astronomer, with no restriction on nationality, chosen by a Search Committee on an international basis, and appointed initially for a fixed term. It was proposed that the Search Committee, chaired by the Chairman of the Board and selected by the Board, should contain one member of the ANU among its two Australian members, with two British and two from elsewhere. The Board hoped the Committee could report in January 1973. In specifying the duties of the director, it saw as one of his first tasks the submission of advice about the numbers and nature of the scientific and technical support staff, which he considered necessary. At this time, the Board was thinking in terms of about three or four astronomers, probably on secondment for a few years, and the direct recruitment of about four specialist technical staff. The objective would be to build up the team over two years to match the programme of assembly and commissioning.

A Committee of the University Council considered the Board's views. The reaction was less than wholehearted acceptance, although the Board thought that some grounds for cautious optimism could be discerned in the response. The ANU was ready to accept in principle the intentions of the Board to appoint a director and to set up a Search Committee to seek the most suitable candidate. However, the ANU could not agree to the proposals for supporting staff without further information as to how such staff would fit in with the staff already at Siding Spring, and an assessment of how much support would be needed from ANU staff and facilities on the mountain. This attitude contained the seeds of yet another deadlock between the Board and the ANU; a first-class candidate would be unlikely to accept the directorship as long as outstanding issues separated the two groups, but until the new director had been designated and consulted, the Board was not prepared to commit itself further on the matter of support staff, which now remained the main point at issue. Nevertheless, the Board was sufficiently confident after the meeting to release a press statement on 11 October on the progress in negotiation. The fact that the University did not agree to the statement illustrates the rearguard action to which Eggen, at least, was still committed.

On 12 October, Hohnen, the ANU Secretary, drew the attention of the Minister to a very rough estimate for upgrading the AAT and providing additional instrumentation which had appeared in a Board document prepared by Hosie concerning future financial needs. The figure of $1.25 million per year was a revised version of an earlier estimate of $1 million, which had itself been increased on Hosie's initiative from a previous Australian estimate of $600,000. This lower figure for future running costs had contained no allowance for upgrading and instrumentation, and without such continuing investment, the performance of the AAT relative to other telescopes

would have soon declined. It was the querying of items such as this at the highest
levels that continued to cast a shadow over the project.

8.10 Change of Government

Despite this, the Board members felt that the worst lay behind them. They could
concentrate now on the important tasks of the assembly of the telescope and the
search for a director. But the respite was illusory. A second crisis arose at the end
of 1972. In December there was an election in Australia which for the first time
in decades brought to power a Labor Government with Gough Whitlam, the new
Prime Minister. The new Minister for Science was W.L. Morrison, who had been
a diplomat before entering politics in 1969. The stage was now set for Crawford
to make one last stand against the Board. On 20 December, the day after the full
Whitlam Government had been sworn into office, H.C. Coombs, Chancellor of the
ANU and a man with as many friends in the right places as Crawford, wrote briefly to
the Prime Minister. Addressing the decision of the Board to appoint its own director
and staff, and arguing that the Board's proposal involved a waste of technical and
other resources, he concluded that '. . . *It is desirable that this matter should be put
on the right lines as soon as possible.*' On 18 January 1973 Crawford and Coombs
met with the Prime Minister, who handed the matter to Morrison with instructions
to observe the spirit as well as the letter of the Agreement.

At about this time, Sir James Price, then Chairman of CSIRO, suggested that
Taffy Bowen should become Science Counsellor at the Australian Embassy in Washington for a three-year term. For Bowen, a retired Chief of the CSIRO Division of
Radiophysics, this was a highly attractive position. It offered a useful salary above
his pension and made him unusually well placed to keep in touch with colleagues in
the American scientific and military establishment with whom he had worked closely
during the war years. Bowen agreed, provided that CSIRO would consent to his
continuing as Board Chairman. Price concurred. The Department of Science also
agreed, and Ennor saw advantages in having the Board Chairman in Washington
seated midway between the two countries.

Shortly after the Labor Government was in office, the Board Chairman was sum-
moned to appear before the new Minister for Science. Bowen relates how Morrison,
the former diplomat, proceeded to dictate what course the Board would take over
the management of the telescope. Bowen advised that the Minister had no responsibility for the matter, and in law the responsibility was vested entirely in the Board.
He pointed out that the Board was a properly constituted body, established by two
Governments, whose responsibilities were clearly defined by an Act of the Australian
Parliament, promulgated several years ago. Morrison announced that his Government would not agree to any arrangement other than that the ANU would run the
telescope. Bowen describes how the Minister's argument continued along lines some-

what removed from constitutional practice. When informed that he might be relieved as Chairman, he advised that a Board member could not be removed on the whim of a Minister; Australian legislation also dealt clearly with such events. Morrison finally produced his ultimate weapon.[7] He pronounced that Bowen could not take up an appointment in Washington and at the same time remain Chairman of the Board, and that he needed to be close by so the Chairman and the Minister could consult on a daily basis. Bowen quickly weighed up the attractions of the Washington appointment with its certainty of a substantial salary. He had been fighting the AAT's battle for three years; it was enough. He told Morrison he would resign as Chairman of the Board and the interview was over.

Unknown to Morrison, the Board's rules on the Chairmanship prescribed that the next Chairman would be from Britain, and Sir Fred Hoyle, based in England was appointed. Whatever Morrison's motives might have been, popular perception of the event was that although the ANU had lost the telescope, 'they' at least 'got' Taffy.

In a letter dated 13 February 1973, the Prime Minister suggested to his Science Minister that Morrison replace Bowen with a senior member of the ANU after he had resigned to take up his Washington appointment two months later. But the important matter of the AAT management still remained unresolved. There was frantic briefing of the new Minister by his department and by 31 March J.P. Lonergan, then Acting Secretary of the Department of Science, recorded that in reference to Coombs' note to the Prime Minister, he (Lonergan) '. . . *was left in little doubt that, in some important respects, Dr Coombs' note does not portray a balanced view of the problem'*.

On 2 February, the ANU's Research School of Physical Sciences held its first Faculty Board meeting for 1973. This was an illuminating episode on several levels. The principal item discussed was a proposal that the University establish an Observatory Services Unit (OSU) which would provide support facilities for the telescopes at Siding Spring, notably for 'foreign' telescopes like the AAT and the UK Schmidt. The Unit would be administered by the Director of Mount Stromlo Observatory, but responsible to the Vice-Chancellor and a committee appointed by the Council of the University. The minutes record that this meeting was asked to review an earlier decision on the matter, '. . . *because that decision had been based largely on Professor Eggen's statements that he would be Director of the 150-inch Telescope and would control all activities at Siding Spring; statements now known to be incorrect. . .*' Sir Ernest Titterton, the Director of the School, commented that '. . . *several of Professor Eggen's statements were misleading. In particular, the representation of the situation* [regarding management of the AAT] *as a struggle between British and Australian astronomers was not a true picture. . . Regrettably, the real situation was developing into a struggle between the ANU* [on the one hand] *and the Australian and British astronomical communities* [on the other].' That Eggen regarded the ANU as a principal party to the AAT Agreement is evident from the minutes.[8] The Director of the School saw the Agreement differently, describing it as '. . . *an agreement between the UK and Australian Governments. The ANU interest was, in a sense, peripheral although it, like other astronomical groups, had a vital interest in access to the AAT for its astronomers'*. This view certainly seems the more reasonable. The Faculty Board reversed its former decision, which favoured setting up the OSU, but the University proceeded with it anyway. The OSU remained in existence until Eggen resigned on 30 September 1977, after which it was dismantled.

8.11 Towards a resolution

The AAT Board held its fifth meeting in Canberra in late March 1973. The Search Committee had applied itself assiduously to the task of selecting possible candidates for the directorship, and invited two candidates to meet the Board in Canberra. It was unfortunate that they both felt unable to accept the offer of the position, and it was evident that uncertainty over the financial policy and over the future relations with the Mount Stromlo astronomers played a significant part in their decisions. The Board, therefore, deferred the appointment of a director in the hope that the remaining policy uncertainties would soon be resolved.

A further effort was made to clarify, if not resolve, the problems between the Board and the ANU through meetings between the Board Chairman, Bowen, ANU representatives and Ennor in early April. Nothing was solved, but differences were brought into sharper focus. However, a letter to Morrison would shortly precipitate a final resolution. On 11 April, R. Hanbury Brown, Professor of Physics (Astronomy) at the University of Sydney, wrote to the Minister. He attributed much of the trouble with the ANU to the fact that Australia lacked a satisfactory policy for the provision of large-scale scientific instruments available to the whole community, and that the ANU had no claim on the AAT since Mount Stromlo neither spoke as a national facility nor aimed to become one. Hanbury Brown had first-hand knowledge of how a national facility operated as a result of his association with the National Radio Astronomy Observatory at Greenbank, West Virginia, where he had been a member of the Board of Visitors. In the early 1970s, he was probably one of the few people in Australia who appreciated the concept of a national astronomical facility pioneered in the USA. Eggen, he claimed, was the only university representative on the AAT Board, yet he had not consulted other Australian universities on any matters concerning the telescope. Finally, he warned that the Board would not attract a good director for the AAT, as long as harmonious operations at Siding Spring depended on personalities. His sentiments clearly reflected those of the state universities which expected equal access to the new telescope with ANU astronomers. Eggen, however, did not believe in access to observing time *as of right* but only *by scientific merit*. On a matter of principle, Hanbury Brown's claim for a national facility was justifiable, but in practice astronomers at Sydney University and elsewhere in Australia always had fared well in obtaining time on Mount Stromlo telescopes. It was generally acknowledged that in this respect Mount Stromlo Observatory's record was good.

Morrison organised a meeting of senior Australian astronomers in Canberra on 12 May 1973, which basically went over the same ground covered by Ennor at his meeting in July 1972. Again, an overwhelming and long-patient majority voiced its support for an AAT with a director and staff employed by the Board. Morrison advised the Prime Minister that the Government should support the Board, but without any commitment to the staffing levels and future financing. Finally on 12 June 1973, the Australian Government officially endorsed the Board's decision to appoint its own director and staff.

A number of important changes had taken place earlier in March 1973. The fifth Board meeting held that month marked the end of membership on the Board for Taffy Bowen, Ken Jones and Jim Hosie. We have narrated already the circumstances of Bowen's resignation. He was replaced by J.P. Wild, Chief of the CSIRO Division of Radiophysics, the same post Bowen himself had held for 25 years. Jones resigned when he was appointed Permanent Head of the Department of Education. Under the new Whitlam Government, the Department of Science and Education had become two separate departments. Sir Hugh Ennor, the Permanent Head in the combined department, remained with Science, while Jones, his First Assistant Secretary, moved to Education. On account of the problems between the Board and the ANU, Morrision's instructions were that the Permanent Head in future must sit on the Board; so Ennor became the next Board member.

Ennor's appointment was significant. Throughout the troubles with the University, he had supported the Board's view on management, notwithstanding his previous close ties to the ANU. He had been Dean of the John Curtin School of Medical Research, and had acted as Deputy Vice-Chancellor during Huxley's term before joining the then Department of Education and Science. He believed that it was in the best interests of astronomers outside the ANU if the AAT were independently operated, and he was always mindful that the Australian view – whatever it was – was only half of the matter. Nevertheless, as early as 25 February 1972, Ennor advised Jones that by virtue of the fifty–fifty arrangement between the British and Australian Governments, there could scarcely be any doubt as to who played the dominant role (the Board) and the subordinate role (ANU), even if the latter were dominant among subordinates, and since Mount Stromlo Observatory was unquestionably the dominant force in Australian optical astronomy at this time, and the University could have expected to figure prominently in any negotiation for a large telescope.

Ennor had a difficult task both in obtaining and implementing the views of the Australian community on the management of the AAT. There was no natural forum in Australia, as existed in the British system, for astronomers to voice their wishes.

There had been a general meeting in Australia before the first major conference on AAT instrumentation in August 1968 at Cambridge. But it was not until July 1972, when Ennor organised the first meeting of astronomers, that he had the opportunity to hear what they wanted. His efforts to implement these views met with political forces which were not necessarily obvious to outsiders. At this time, the Australian Liberal Government was in crisis and would be replaced by the Whitlam Labor Government before the year was out. The more the discussion of the AAT management was prolonged, the more acutely Ennor was aware of the debate which had taken place among Canadian scientists in 1968 over the location of the large QEII optical telescope. In particular, he was aware that Prime Minister Trudeau resolved the conflict by cancelling the whole project. He did not want the uncertainty over the AAT management to precipitate a similar resolution by Fraser, or a withdrawal from the agreement by the British. Yet the last days of the Liberal Government presented the least desirable environment in which to argue the views of the scientific communities on the AAT management with Fraser, who continued to support the Crawford line. However, if Ennor had difficulties with Fraser, worse was to come in his dealings with Morrison, the Labor Minister for Science. It is well-known that a very uneasy relationship existed between the Department of Science and Morrison, so Ennor had to wait a further six months before Morrison could be persuaded to endorse the Board's decision on an independently operated telescope.

In Britain, Jim Hosie was transferred to the post of Director, Finance and Administration in the Science Research Council; he was replaced on the Board by M.O. Robins, Director of the Astronomy, Space and Radio, and the Science Divisions of the SRC. The departures of Bowen, Jones and Hosie deprived the Board of a wealth of experience. All three had worked tirelessly to establish the AAT on a sound basis, and it was ironic that so soon after their departure from the Board, the major policy issue on which so much time and effort had been spent, was to be settled. Moreover, the Board was about to enter the final phases of assembly and commissioning of the telescope, and it was no closer to having a director.

Also in March 1973, Sir John Crawford reached the end of his term as Vice-Chancellor of the ANU and was replaced by Professor R.M. Williams. This change was a significant step towards easing the tension which had grown between the Board and the University. The next Board meeting in July was also the last meeting for Olin Eggen, who resigned as a member in August. On his departure, he wrote to Williams that he saw the Board's decision on the AAT management as creating government astronomy at the expense of university astronomy.

It is interesting to reflect on the course events might have taken if Bart Bok had been Director of Mount Stromlo Observatory instead of Eggen, or if Huxley had been ANU Vice-Chancellor instead of Crawford during these stormy years. Bok appreciated that the operation of the AAT would be of such a size and complexity that it would best be carried out by its own independent managers. Huxley had often told Gascoigne that he did not want the University to accept further astronomical responsibilities in addition to those it had at Mount Stromlo and Siding Spring. By the end of the decade the astronomical staff at Mount Stromlo seemed unanimous in the view that it was just as well they had not had to shoulder the formidable burden of operating the AAT. Instead they were free to pursue their true vocations, which were using it to do astronomy, and training graduates. Paul Wild's words at the meeting called by Ennor in July 1972 had come home to roost: *'Tell us, Olin, why exactly do you want to take on all these extra chores?'*

Notes to Chapter 8

1 See Appendix 2
2 Titterton 11 May 1970
3 Correspondence quoted in this chapter is found in the archives of the Australian Government agency.
4 The page bearing this extract in the ANU had been endorsed *Confidential – not to be quoted without Vice-Chancellor's consent.* (Department of Education and Science archives AAT 71/368)
5 Burbidge G. (1972). *Nature* (London) **239**, 118
6 See Chapter 5
7 Hoyle F. (1982). *The Anglo-Australian Telescope*, University College Cardiff Press p.19
8 Faculty Board Minutes 2 February 1973

9   The beginnings of the Observatory

9.1 The first staff and scientific base

The uncertainty about the Board's controlling its own staff to manage and operate
the telescope was finally resolved by the decision of W.L. Morrison, the Australian
Minister for Science, in June 1973. Shortly before, the ANU tried to meet some of the
criticisms of the Board that too much power would be vested in the Director of Mount
Stromlo Observatory if that Observatory were responsible for the management and
operation. It therefore proposed establishing an Observatory Services Unit, separate
from the Department of Astronomy and the Research School of Physical Sciences.
We referred to this in the previous chapter. However by June, events had moved too
far and the Board was convinced by experience that it must have direct control of
the crucial aspects of management and operation.

When the Board met in London in July 1973, it had a clear mandate for the first
time to appoint its own staff to manage and develop the AAT. As well, it kept in mind
Morrison's wish that the outstanding problems between the Board and the University
had to be settled. This meeting saw major changes in the Board membership with
Fred Hoyle, now Chairman, and three new members, two of whom were representing
the contracting parties: Ennor was now the Australian Government representative,
and on the British side M.O. Robins replaced Hosie. The third new arrival was J.P.
Wild, replacing Bowen. It was also the last meeting for Olin Eggen whose resignation
occurred soon after.

Two important matters occupied the Board members for many months to come.
The first was relatively straightforward: how the Board's employees should have
right of access to those services outside the AAT dome to which the Board was
contributing 60 per cent of the costs. The most important service was in the Utilities
Building, housing machine tools for the use of both Board and ANU staff. In the
prevailing climate of difficulty and uncertainty, the Board held the opinion that the
preferred solution would be to negotiate to buy out the University's 40 per cent share.
Arrangements might then be made for ANU staff to use the facilities under contract.

The second matter was of more far-reaching importance: the location of the scientific base essential for an independent AAT Director and staff. As the location of
this base would be of special interest to visiting astronomers from both Australia and
Britain, their views were sought before the Board made a firm decision. The general
feeling was that a base adjacent to Mount Stromlo was preferable to allow sharing
of facilities as much as possible, provided the arrangements were acceptable to the
University and the Board. Alternatively, a site adjacent to the ANU Department of

Figure 9.1   Board member (1973–78) M.O. Robins, CBE

Physics seemed attractive. Among other possibilities was a site in Sydney convenient to the University of Sydney or to the CSIRO Radiophysics Laboratory.

The next move took place in August when the Board complying with Morrison's request met ANU representatives in Canberra. In the meantime, Eggen had resigned from the Board, but now sat on the ANU side of the negotiating table opposite the Board. There was a frank exchange of views, the University stressing its perceived rights and obligations under Article 4 of the Agreement. The relevant part of this Article is clause (2) which states

> So far as practicable and subject to satisfactory arrangements being made between the Telescope Board and the University, use should be made of supporting facilities in existence or to be provided by the University at Siding Spring Mountain and at Mount Stromlo. This does not however preclude the use of supporting facilities elsewhere.

The Board expressed its concern over difficulties its employees were experiencing in obtaining access to the Utilities Building and in using the machine tools. In view of the University's firm assurances to remove these problems, and to have rules for the use of the building agreed between the AAT Project Manager and the ANU Mountain Superintendent, the Board withdrew its proposal to buy out the University's 40 per cent share of the building. The Board outlined the needs for a scientific base for its operations; and preliminary studies indicated that some 10,000 square feet would be required to accommodate 25 to 30 persons including the Director and four to six

Figure 9.2   The AAT dome and the utilities building

scientific staff. First thoughts that it might be feasible to have such a base at Mount Stromlo did not survive subsequent discussion with the University, and the Board realised that pursuing the matter further would be fruitless.

A small working party which had been considering the feasibility of several possible sites for the scientific base presented its report at the Board meeting in November. The members of this group[1] had assumed that the main tasks to be carried out at the base would be the development and trials of new instruments, and the control and administration of the whole facility. They had visited and discussed the matter at the CSIRO Division of Radiophysics in the Sydney suburb of Epping, at Mount Stromlo Observatory, at Coonabarabran and at Siding Spring Observatory. The possible sites were within the complexes at Mount Stromlo Observatory, at CSIRO Epping, on Siding Spring Mountain, within the AAT building and in or near Coonabarabran. A base within the AAT building, though feasible, was ruled out as technically undesirable. The town of Coonabarabran was not favoured because it was too far from other scientific institutions; nor would Siding Spring provide a suitable scientific environment for a base. There were no technical objections to Mount Stromlo Observatory, Epping and Siding Spring. Therefore, the working party recommended a choice based on administrative and financial grounds.

The Board realised that the choice of a permanent site for the base was likely to be difficult because it was already clear that many Australian astronomers were strongly opposed to a site away from major centres of activity, which means the capital cities

Figure 9.3   The AAO laboratory at Epping

in Australia. They disliked Siding Spring and Coonabarabran for this reason, and believed it would be difficult to attract and to retain highly qualified technical staff in the relatively remote locations of the telescope and of Coonabarabran. Most British astronomers, eager to make their long journeys to the AAT as effective as possible, believed that the scientific base should be as close to the telescope as practicable – if not on the mountain, then at Coonabarabran. The fact that the first Director whose views on the site would be important had not yet been appointed, and that the first two years of operations would require the Director and his staff to spend much time at the telescope persuaded the Board of the merits of a temporary solution to the site problem. This also would enable the choice of a permanent site to be based on some experience when all factors would be clearer. It was decided to make use where possible of accommodation in the telescope building, and to seek agreement from CSIRO to locate temporary office buildings within the grounds of its Epping Division and to have temporary use of the workshop facilities and library there. In addition, the Board arranged to have the use of an office at Mount Stromlo Observatory.

For some time the selection of the first Director was a source of concern to the Board. Taffy Bowen commenced a personal search for a suitable person but found that the troubles with the University over the control of the telescope had seriously tarnished the appeal of the position. Hoyle's own search in America confirmed this. Nevertheless by March 1973 the Search Committee, described in Chapter 8, had found two candidates who were invited to be interviewed in Canberra. These were R. Hanbury Brown from the University of Sydney and J.B. Oke from California, but both rejected the offer of the Directorship because of the continuing unsettled nature of relations between the Board and the ANU.

Once the uncertainty was removed about the precise responsibilities of the post, particularly vis-à-vis Mount Stromlo Observatory, the Board offered an appointment for two years to E.J. Wampler of the Lick Observatory in California. An American astronomer and a leading instrumentalist of his time, Joe Wampler was well qualified to head the teams soon to assemble at Epping and Siding Spring. With his fellow American, Lloyd Robinson, Wampler had built a very sensitive photon detector, the Image Dissector Scanner, which they had used very successfully on the Lick 120-inch telescope. In accepting the Directorship, he brought to Australia his expertise and his colleague to build a similar instrument for the AAT. This expertise came at a critical time when the new telescope needed a state-of-the-art instrument to get itself going.

It now became urgent for the Board to define a firm policy for the appointment of scientific staff. This policy would have to meet the needs of the Director in the development and operation of the telescope and necessary instrumentation, and also be acceptable to the Australian and British astronomical communities. This posed the problem common to most large facilities intended to be used by scientists based in universities or otherwise not staff members of the facilities: how to ensure that the facility would be adequately staffed by able people, who would undoubtedly demand a share in the use of the facility, benefiting from their inside knowledge, without putting the outside users at a disadvantage. Staff at the Kitt Peak National Observatory received 40 per cent and visitors 60 per cent of the observing time.

British astronomers were particularly sensitive on this point; they were apprehensive lest astronomical staff based at the telescope should monopolise the available time or take undue advantage of the instrumental facilities to the detriment of British based users. On the other hand, it was evident on closer consideration that the AAT with its advanced control and instrumentation systems simply could not be operated by visiting astronomers from anywhere unless there was a core of experienced astronomers attached to the AAT able to provide continuity of support and advice to visitors. Such a core, including the Director, needed an assurance of reasonable observing time to attract and to retain staff of sufficiently high calibre. The AAT Agreement required that observing time be divided equally between the Australian and British communities, and it was the Board's policy that the two Governments should each allocate its share of time to applicants as it chose. It was inevitable that the policies on scientific staffing of the AAT and the systems of time allocation should become independent. The problem facing the Board was how best to reconcile these requirements.

The Board considered it essential that the Director should have a staff of six astronomers whose interests encompassed both observations and instrumental developments. For continuity, the appointments would be of varying length up to five years. It was suggested that there should be an initial allocation of observing time and, after a period, the applicant would need to seek more time in open competition through established channels. Visiting astronomers would be supported by technical staff in the operation of the telescope and be advised by the scientific staff on the functional characteristics of instruments. The latter also would offer limited collaboration to visitors. The Science Research Council was generally in agreement but emphasised that it believed the Board's scientific staff should be concerned mainly with programmes of collaboration with visitors, and that instrument development by them should be on a modest scale, again with the astronomers of the two countries playing a substantial role.

The period from April until the inauguration of the AAT in October 1974 was one of much administrative activity, in addition to the great engineering efforts being made to bring the telescope to a state of readiness for commissioning. In agreement with the Department of Science and the CSIRO, the Board made plans to commence operations at Epping in September in three prefabricated huts. The first staff to take up residence here included the nucleus of an electronics section, established by Lloyd Robinson during his visit from the Lick Observatory. When the Project Office was closed down in Canberra in 1975 a mechanical and optical design group was brought together at Epping with a small administrative staff. At Siding Spring there were still difficulties over the use of the Utilities Building, and a further Working Party was established to investigate the problems in detail.

At the meeting in London in August 1974, two new members joined the Board: Professor R. Street and Professor V.C. Reddish. Street, who replaced Olin Eggen, was the Director of the Research School of Physical Sciences at the ANU, and his appointment was very important for resolving the problems between the Board and the University. Vincent Reddish of the Royal Observatory, Edinburgh, was responsible for the Schmidt Telescope, which was owned and operated at Siding Spring Observatory by the SRC. Reddish replaced Margaret Burbidge when she relinquished her post as Director of the Royal Greenwich Observatory. It was fortunate that the two new members brought first-hand experience of the two organisations so closely linked to the AAT operations.

The arrival of Joe Wampler in September 1974 gave a fresh impetus to the search for solutions to the problems of future management arrangement, the location of the scientific base and the levels of staffing for the operational phase of the telescope. On management arrangements he had found no consensus view from other expensive government-funded facilities, such as ESO, and he believed that the AAT would best serve astronomy if freed from entanglements with other organisations. The telescope should be operated as an observatory with a strong rotating scientific staff, whose members should attract sufficient observing time to enable them to pursue important and difficult research programmes. This staff should also assist visiting astronomers. Wampler thought about 20 nights of observing time per year per staff astronomer might be appropriate. On the question of the choice of site for the scientific base, his first concern was that the location and nature of the buildings must be decided soon, since the need for access to dark rooms, measuring equipment, machine tools and a library was becoming ever more urgent. He did not favour too close a union with Mount Stromlo Observatory or indeed any other institution, and regarded Coonabarabran as too isolated, but believed that there was little to choose between Canberra and Sydney, each offering proximity to academic communities.

After considering the Director's views on the management arrangements, the Board generally agreed the basis of a plan. There should be 35 nights per year allocated at the discretion of the Director, and the remaining 330 nights divided equally between Australian astronomers, British astronomers and the AAT scientific staff and the Director. Such a plan would mean each Government would contribute ten nights to the Director, 54 to AAT staff members and 118 to its home-based users. At about this time, the Board advertised six scientific positions, and mid-1975 saw the arrival of the first astronomers on the staff. These were I.J. Danziger, P.G. Murdin, M.V. Penston, B.A. Peterson, B.L. Webster and J.A. Whelan. In addition, the first Fellows funded by the SRC, D.A. Allen and R.A. Fosbury, joined the group at Epping.

The Board found greater difficulty in considering the site for the permanent base, and would have preferred to defer this matter until it had experience from operating the temporary base at Epping. But, as the Director firmly pointed out, uncertainty about the final stages was causing many difficulties in recruiting staff. Although the temporary arrangements with CSIRO at Epping could satisfy short-term needs, the growth of the whole establishment was being increasingly hampered by this uncertainty. By now the possible sites had been reduced to three, but the Board found it impossible to agree on either Epping, Canberra or Coonabarabran, partly because the relative costs had not been established in sufficient detail. A more detailed costing exercise was undertaken before a decision could be made. The Board convened in San Francisco in December specifically to decide where to locate the permanent scientific base. Yet, after prolonged and intense discussions, it failed to find an agreed solution, with views split along national lines; Hoyle, Reddish and Robins held one view, Ennor, Street and Wild another. This split represented to a large extent the different attitude towards and the understanding of social factors in Australia arising from the size and remoteness of the country.

The Australian members were very firm in their belief that many scientific advantages could be gained by the association of the Board's main activities with an Australian university or research institute. Experience in Australia of the operation of scientific projects showed that they were better organised as field stations with headquarters at a home base. For the AAT they favoured either a Sydney or a Canberra base. They believed Coonabarabran presented serious difficulties in recruiting suitable staff; experience proved that there would be a high turn-over of staff, particularly in the technical grades, in rural New South Wales. The Australian members also believed that advanced technological developments would be significantly more viable in a large city, and that the professional health of the staff would suffer if isolated from an academic environment and from industrial contacts, which was inevitable at Coonabarabran.

The British members looked at the problem from the point of view of the most efficient management of the telescope. Ideally, the scientific and support staff and the operational and maintenance staff of a central research facility should be located together at the facility. This would give maximum opportunity for the interchange of ideas both internally and with the stream of visitors coming to use the facility for either short or long periods. Since a location adjacent to the telescope would not be feasible, the British members favoured Coonabarabran where AAT users from Britain would most effectively and conveniently meet both the scientific and the operational staff. If the base were elsewhere, additional journeys would be needed and visitors would be inclined to by-pass such a base to go directly to the telescope. The British members agreed that while recruitment problems would exist at Coonabarabran, these would be temporary and that dedicated astronomers who would comprise the scientific staff may wish to be located as near as possible to the telescope.

In the circumstances, the Board Chairman, Fred Hoyle, did not put the question to a formal vote. Instead, the Board agreed to continue for a time the temporary arrangements at Epping, and to reconsider the situation with the respective Government agencies. The additional costs and uncertainties resulting from this delay added to the urgency of finding an acceptable solution.

Among other matters considered at San Francisco was the question of the total staff complement needed for efficient operation of the telescope, together with the supporting maintenance and development activities. Governments have always been sensitive to the growth of staff numbers in government-funded institutions, and the AAT was no exception. The Director's bid for a total of 58 at this stage was given a preliminary examination, and the Board's view was that 50 would probably suffice, subject to further consideration when more experience had been gained. It was also at this time that the Board decided to separate the functions of the Board Secretary from those of its Administrative Officer on the staff responsible to the Director. Both duties had been carried out with distinction since the very early days of the project by Doug Cunliffe. Taffy Bowen seconded him to the Project Office in 1968 from the CSIRO Division of Mechanical Engineering, and Cunliffe became the most senior non-technical person on the Observatory staff. With his characteristic unfailing good humour, he continued to serve as Executive Officer when the telescope moved into the operational phase with the temporary base in Epping. From that time the Secretary of the Board was an officer on secondment from the Department of Science who remained in Canberra, thereby satisfying Article 10 of the AAT Agreement, which stated that the principal office of the Board would be in the Australian Capital Territory.[2]

The San Francisco meeting achieved little, and the Director was left with unresolved problems on the location of the permanent base and its repercussions on staff recruitment and growth of the facilities, and about the staff complement. He was unhappy about relations with the ANU and the sharing of facilities with Mount Stromlo Observatory staff on Siding Spring Mountain, on the lack of firm information about the operational budget within which he would have to work, and he needed advice on administrative staffing matters such as the terms and conditions of employment to be offered staff. In an effort to clear away at least some of these difficulties without waiting for a full Board meeting, it was agreed that the Secretary of the SRC and Mac Robins should visit Australia in February 1975 for informal discussions with the Director and the Australian members of the Board. The SRC Secretary at this time was R. St J. Walker; he had not visited Siding Spring, and it was arranged that he should do so with Wampler and Robins.

It was usual to fly from Sydney to Coonabarabran in a chartered light aircraft. On this occasion the flight ended shortly after take-off with an emergency landing at Bankstown Airport in the west of Sydney, where the passengers were stranded for several hours in an aircraft maintenance hangar. Here a draft document setting out proposals for the operations of the AAT was hammered out with the Director for discussion with Board members. This became known as the *Bankstown Manifesto*. There was wide-ranging discussion about the Director's problems of working within the necessary administrative framework, for Wampler had a true American disregard for red-tape, and found it difficult to deal with the Board's administrative system. However, he was reassured of the Board's support, within available resources, for his aspirations to operate the AAT with a high degree of efficiency and to develop the facilities for the benefit of all users. The *Bankstown Manifesto* was accepted with minor changes by the Board and issued as a policy statement in February 1975.

In the debate with the ANU there was the continually repeated claim by the University that the AAT Agreement permitted the Board to run a telescope, not an observatory. In the *Bankstown Manifesto* the Board ultimately made it clear that it was quite prepared to run an observatory, and this it called the Anglo-Australian Observatory which would have its temporary administrative base on CSIRO property at Epping, in Sydney. In addition, the document affirmed that as far as practicable the Board would make arrangements for sharing facilities with the University. It emphasised the principal role of the Observatory as a service facility, and the Board's right to make rules governing how it would operate the telescope. It clearly stated that staff would be appointed on short-term contracts, and would have to compete for observing time with other British and Australian astronomers.

The evident welcome given to the young organisation by the CSIRO staff of the Radiophysics Division was a promising start to establishing the AAO's temporary base at Epping. This encouraged the Board at its meeting in Canberra in May to leave the base there at least until the end of 1978. During this time, the Board was fortunate in being able to purchase the library from the Radcliffe Observatory in Pretoria when it closed in 1974.[3] The Radcliffe Collection has been a most valuable asset for the Anglo-Australian Observatory, and is a reminder of the long Anglo connection with southern hemisphere astronomy. Its arrival at Epping when the Observatory was settling in with the radio astronomers was timely. However, relations between AAT and ANU staff at Siding Spring were less settled. There was much friction and possible misunderstanding about the freedom of access to use the equipment in the Utilities Building, and yet another Working Party was set up to sort out the problems once and for all.

The Board meeting in May 1975 was the last for Hoyle. He had been associated with the project since 1968 and, contrary to his original intention, had remained on the Board and become Chairman in 1973 after Taffy Bowen's departure to steer the project through some very difficult times. Throughout the years of turmoil he had steadfastly and successfully maintained the scientific objectives of the project, and had contributed greatly to the operational success soon to be achieved. His place as a British Board member was taken by Professor Sir Harrie Massey, and Hoyle left the next Chairman, Paul Wild, with some challenging problems to solve.

The longer-term development of the Observatory and of the site for a permanent base continued to occupy the Board at its Edinburgh meeting in October that year. Massey was more aware of the sensitivities in Australia than any of the other British Board members, and agreed with Street to delay the choice of a permanent site until the concept of the Anglo-Australian Observatory had been further developed. Reddish, on the other hand, believed that astronomers would gravitate to the site of the telescope, and if the Board had facilities there and at, say, Sydney, the Director could make his own base at either place. With respect to the longer-term development of the Observatory, the Board recognised that the existence of three separate astronomical groups at Siding Spring – the AAT, the UK Schmidt Telescope, the ANU telescopes – invited questions about the desirability of some form of combined policy direction.

However, Wampler's impending departure after completing his two-year appointment prevented any immediate changes. The Board, in fact, decided to re-affirm its policy of maintaining direct control of the AAO. There was also a strong feeling that all users of the AAT should be encouraged to initiate development of instruments where appropriate, and should not rely solely on the resident staff. It was apparent that a strong in-house scientific and technical competence would have to be preserved for the operation and maintenance of the telescope. It was in the pooling of engineering and technical support with the other groups on the mountain that the Board saw the most likely opportunities for economy and efficiency. The continued pressure from the two Government agencies for financial economies and for the absolute minimum growth in staff numbers led the Board to give first priority to consolidation and maintenance of a high standard of reliability.

Wampler resigned and returned to the Lick Observatory at the end of March 1976, having set the AAT on a good instrumental base with the photon counting device. In the interim period before the arrival of his successor, Paul Wild was Acting Director, and Doug Cunliffe and Peter Gillingham took executive control at Epping and Siding Spring respectively. Although much remained to be done in the way of instrumentation and the full exploitation of the excellent engineering features of the telescope, the political and administrative aspects of the project were beginning to settle down. There would continue to be the annual struggle to achieve adequate financial contributions from the two Government agencies to support what the Board believed to be necessary developments, but the main outstanding question was the permanent site for the scientific base.

At its next meeting, the Board again addressed this question. However, the difference of opinion between the British and Australian Board members was a matter of great concern to the smooth running of the AAT. Before the meeting Sir Harrie Massey discussed in depth with staff at Epping and Siding Spring the various sites for the base, and came to the conclusion that the Laboratory had to be in an urban location. To the relief of the Chairman, Paul Wild, Massey carried his British colleagues with him and must take full credit for resolving the impasse. There was unanimous agreement that the Board should maintain a permanent laboratory at Epping. The British side, in accepting that the people in Australia would have to handle the detailed day-to-day problems of the operation, ultimately respected the local wishes.

The decision recognised two essential requirements for the permanent site, that, on the one hand, the environment should encourage interaction between the AAO staff and other astronomers and engineers, and provide living conditions for the staff and their families; and, on the other, that the telescope itself should become the focal point of the Observatory's scientific and technical facilities. The Board agreed to encourage the closest possible liaison between the staff based at the telescope and those at the laboratory through frequent interchange and visits of staff and improved communications. It was recognised that there should be a continuous scientific presence at the telescope, and it was hoped that visiting astronomers would spend some time at each location. When the new Director took up his post, he would have a well-defined framework in which to work, without the uncertainties surrounding a temporary base, and would be encouraged to see the Laboratory and the Telescope as complementary parts of the Observatory, neither being recognised as the headquarters.

In 1985, in the context of a British review of its involvement in southern hemisphere astronomy, the Board studied the benefits of moving the Epping base to Coonabarabran. Results showed that there was little to be gained financially by moving the staff and operations from Sydney, and in fact the staff was completely in favour of remaining in the city. It was interesting to note that the very people (the astronomers) who in 1974 were expected to be keen to live as close as possible to the telescope voiced, along with the other members of staff, an overwhelming preference to remain in Sydney with its superior opportunities for secondary and tertiary education. In addition, a decade after the initial long debate and now in a less discriminatory society, the staff made the point which seems not to have been given great weight in 1974. This was the lack of employment and career prospects in Coonabarabran for staff members' spouses, many of whom are skilled professionals.

9.2 The inauguration of the AAT

It was clear in the early months of 1974 that, barring accidents, the AAT would be essentially completed later that year, although there would be a continuing programme of instrument and ancillary equipment development. The Board believed that the inauguration of such an important scientific project, and such an outstanding example of co-operation between Australia and Britain, should be marked in an appropriate manner. Two matters had first to be settled: the earliest date to assume that the telescope would be in a working condition and could be demonstrated without fear of mishaps; and the acceptance of the invitation to perform the opening ceremony by an appropriate person.

It was known that the KPNO 150-inch telescope under construction at Cerro Tololo, in Chile, was nearing completion, and there was friendly rivalry between the AAT and the Kitt Peak groups over which telescope would come into operation first. The AAT was ahead by some months when Ben Gascoigne, the Commissioning Astronomer, took the first photograph on the night of 27 April 1974. This milestone enabled the Board to confirm to the Australian Government that the planning for the inauguration ceremony could begin in earnest.

It would obviously be a complex administrative operation, involving the transport of several hundred people to and from a somewhat remote site, and accommodating them in reasonable comfort in a building which was certainly large but was a maze of differing floor levels containing massive and awkwardly shaped equipment. It was fortunate and most appropriate that His Royal Highness Prince Charles agreed to perform the opening ceremony on 16 October 1974. The detailed arrangements, including all those matters inseparable from a royal occasion, gradually took shape, with Doug Cunliffe carrying a lion's share of the responsibility. There was some concern that heavy rain might disrupt arrangements by cutting the road access from Coonabarabran, because at this time there was no bridge over the Castlereagh River on the road to Siding Spring. Those at the Observatory knew well the inconvenience when the causeway was impassable after days of heavy rains. Doug Cunliffe consulted the Army which suggested that Army trucks could be used to convey guests across the river in the event of very heavy rain. The Army Engineers were on manoeuvres in the area at the time of the inauguration, but luckily their heavy vehicles, so indecorous a measure for the occasion, were not needed.

October 1974 was a vintage month for British astronomers. As the Board members assembled at Siding Spring in readiness for the inauguration, the news came that the scientific satellite Ariel 5 carrying astronomical X-ray detectors had been successfully launched from the coast of Kenya. Then it was learned that two distinguished radio astronomers at Cambridge University, Professors Sir Martin Ryle and Anthony Hewish, had been awarded jointly the Nobel Prize for Physics. All was set for an auspicious occasion, but the Siding Spring weather had a nasty surprise in store. The mountain top was lashed by a ferocious gale on the morning of 16 October. An English Professor of physics was heard to remark that it was the first time he had seen horizontal rain since leaving Manchester many years before. As many of the guests were expected to arrive at the small airport of Coonabarabran with little time in reserve, the organisers were concerned that the bad weather would seriously disrupt the programme. Fortunately this did not happen.

No royal occasion takes place without the raising of flags, and at the AAT inauguration the protocol, of course, was observed. Fred Grigg from the AAT, having had some military service experience, was detailed to attend to the flag raising. Protocol demanded that there would be three flags raised at decreasing heights: first the Prince's standard, next the Australian flag, and lowest the University's flag. Ideally, the poles should have been of different heights, but as these could not be arranged Grigg had to set the flags appropriately. In good time before the event, the Australian and the University's flags duly were set (the Prince's standard is unfurled on his arrival). Later to his surprise, Grigg noticed that the ANU flag was fully raised; he corrected it, and kept an eye on the poles until he observed the ANU flag yet again had been raised. A tall man well over six feet, Grigg once more reset the positions, and secured the cords high out of reach of the culprit, who returned again. But Olin Eggen's final attempt was unsuccessful.

A vivid description of the conditions is given by Fred Hoyle in his personal account of the inauguration.[4] At the time, Hoyle was Chairman of the Board, and it was his task to welcome Prince Charles at the entrance of the telescope dome. He writes:

The vehicles arrived at last and there came a dash from the protocol squad to open the car doors. Gough Whitlam the Australian Prime Minister was the first man up. He moved bravely out through the door of the building. Whitlam is a big man and it was not hard for him to avoid being blown away in the wind, but I recall watching in a kind of fascinated horror as his hair instantly stood straight on end. My turn came next. Move forward, bow, shake hands and make a little prepared speech of welcome, I told myself. But every syllable was blown clean away in the roar of the wind, so that I was instantly reduced to the grimaces of primitive man.

Inside the building about 500 invited guests had been seated beneath the dome itself, with the telescope structure looming overhead. The normally austere surroundings had been softened by the spring flowers which decorated the dais, and the waiting guests were entertained by the Sydney Conservatorium String Quartet playing music by Purcell and Dvorak. This not only added lustre and elegance to the occasion but demonstrated the surprisingly good acoustics of the dome. The group of distinguished guests conducted to the platform by Hoyle included, in addition to Prince Charles, the then Prime Minister of Australia, the Honourable E.G. Whitlam, the then Australian Minister for Science, the Honourable W.L. Morrison, the British High Commissioner, His Excellency the Right Honourable Sir Morrice James, the Governor of South Australia, His Excellency Sir Mark Oliphant, and the then New South Wales Minister for Health, the Honourable J. Waddy, representing the Premier. In addition to the members of the Board, two who had contributed enormously to the enterprise were present: E.G. Bowen, the first Chairman of the Joint Policy Committee, and J.F. Hosie. The Chairman of the British Science Research Council, Sir Sam Edwards, and many astronomers from Australia and Britain had come to witness the ceremony and examine for themselves the splendid new facility.

After presentations to Prince Charles, Hoyle gave the opening address. He acknowledged the support given by the respective Governments and paid tribute to the scientists who conceived the project: Sir Mark Oliphant and Bart Bok in Australia and Sir Richard Woolley and Sir Hermann Bondi in Britain. This was followed by short speeches from the British High Commissioner and the Australian Prime Minister before the Chairman invited His Royal Highness to give the inaugural address.[5] The Prince concluded his speech by declaring the telescope to be open, and the dome and telescope were slewed under computer control as a mark of recognition. Peter Gillingham and Maston Beard had entered appropriate settings into the computer, and Gillingham was seated in the AAT control room ready to start the telescope driving on a cue from Herman Wehner who looked down on the speech-making.

As a memento of the occasion, the Chairman of the Board presented Prince Charles with a gift symbolising the joint undertaking of the two Governments. It consists of a

Figure 9.4   Sir Fred Hoyle presenting HRH Prince Charles with a gift at the inauguration of the telescope in October 1974

Figure 9.5   HRH Prince Charles with Sir Fred Hoyle in the dome of the AAT

Figure 9.6   HRH Prince Charles talking with Hoyle and Gascoigne; Robins and the British High Commissioner, Sir Morrice James, are behind them.

sphere representing the earth, truncated at the latitudes of London and Siding Spring Observatory. The symbolic earth is supported by a disc of material cored from the primary mirror of the Anglo-Australian Telescope. At the latitude of Siding Spring Observatory a quartered diffraction grating symbolises the way in which optical phenomena illuminate man's understanding of the Universe. The gift was created by members of staff of the CSIRO Division of Radiophysics.

Prince Charles was then escorted by the Chairman to the tourist's gallery where he unveiled a commemorative plaque which reads:

Anglo-Australian Telescope

This Telescope, a joint undertaking by the Government of the United Kingdom of Great Britain and Northern Ireland and the Government of Australia, was inaugurated by His Royal Highness The Prince Charles on 16th October 1974.

Before joining the guests for a buffet lunch, His Royal Highness and official guests examined the telescope and some of the ancillary equipment, escorted by the Chairman and Board members. After the formal farewells and the departure of the Royal party, the guests left Siding Spring in rather more clement weather than had greeted their arrival a few hours earlier.

Notes to Chapter 9

1 The members were: J.P. Wild, K. McAlister (CSIRO Divisional Engineer), J.V. Dunn (CSIRO Architect) and H. Wehner (AAT Project Manager).

2 Article 10 was amended in 1986 to allow the principal office of the Board to move to the Epping laboratory in Sydney.

3 We refer to the debate about moving this Observatory from Oxford to South Africa in Chapter 2.

4 Hoyle F. (1982). *The Anglo-Australian Telescope*, University College Cardiff Press p.23

5 The text of the address by the Prince and by Whitlam are in Appendix 8.

10   Commissioning

10.1 First stages

When the telescope was handed over by the manufacturer it was by no means working and complete; rather there was a multitude of tasks still to be carried out before routine observing could begin. Individual components and sub-assemblies had been checked as a matter of course, but from now on it was the functioning of the complete integrated system that mattered, and every aspect of it had to be worked over thoroughly. It was a lengthy process. Some problems were straightforward, but more subtle ones could take weeks or months to track down and eliminate. The commissioning of the Palomar Telescope occupied about two years, during which time the outer part of the primary mirror was refigured and the mirror cell completely rebuilt. It was hoped that with the AAT the process would be quicker, and 1 January 1975 was fixed as a target date on which some observing might begin. But the full potential of the new installation did not become apparent until a little later, in March, when the Wampler–Robinson scanner was given its first trial and visiting astronomers could appreciate for the first time what a superb instrument they had. Full-time scheduled observing began on 28 June 1975, though it was about another year before the telescope was effectively trouble free.

Commissioning proper could be said to have begun on 23 April 1974, when after several weeks of acceptance tests the Project Office was satisfied Mitsubishi and Grubb Parsons had fulfilled their parts of the contract, and formally accepted the telescope. However, for the people on site a more significant date was Easter Monday, 15 April. On that day the cover was taken off the bright red case, and there at last was the mirror, one could reach out one's hand and touch it. The lifting gear was hooked to the crane, and quite soon the great mirror was riding up the hatchway, swaying gently as it travelled the 20 metres from the loading bay to the main floor. It was an emotional moment – an old dream come true, a long journey ended.

The mirror was placed in its cell almost at once, centred to about 0.05 mm as planned. Then the protecting sheet of fabric was stripped off, leaving in Redman's words *'an almost perfectly clean mirror, although Gascoigne felt impelled to dab it lovingly here and there with alcohol-moistened cotton wool'*.[1] In fact everything went well; even the prime focus top end, held up in Sydney by a dockers' strike, arrived in time. Cloudy weather delayed proceedings until the night of 24 April; then, through mist and broken cloud, one could at least verify that nothing seemed seriously wrong. The first real observations were made on the nights of 27 and 28 April, the latter with two hours of good clear sky. Peter Gillingham took a polaroid picture of a knife-edge test, gratifyingly similar to those previously obtained in Grubb Parsons'

Figure 10.1   The AAT primary mirror photographed from underneath as it is being lowered from the observing floor to the aluminising room

test tower; a photographic plate was taken to check the adjustment of the polar axis, and Gascoigne obtained a ten-minute exposure on the globular cluster NGC 6266. All this was done with the mirror not yet aluminised; in fact the aluminising plant had run into a long series of delays and did not leave England until mid-year.

One encouraging feature was the stability of the prime focus cage, which seemed almost immune to man-made disturbances. There are telescopes in which the prime focus observer hardly dares to cough, let alone shuffle his feet, for fear of putting the whole top-end into oscillation; this was not true of the AAT. Another feature was the way in which the telescope could be balanced directly from the control desk. One merely had to note whether the declination motor required more current to drive it towards or away from the zenith ('uphill' or 'downhill'), then move the appropriate counterweight, also controllable from the desk, until the currents were equal. Balance in right ascension could be achieved in the same way. It was a minor matter perhaps, but a great advance over struggles on other telescopes involving long ladders and spring balances.

In the early months the sheer volume of work was almost overwhelming. Everything was crying out to be done, everyone had good reason for wanting the telescope, and for every problem solved two more seemed to surface. The main problem was shortage of manpower. Herman Wehner now had been Project Manager for some months, and both he and Doug Cunliffe had commitments which kept them in Canberra much of

Figure 10.2 The AAT's aluminising plant at the manufacturer's works in England *(Photo: Edwards High Vacuum Limited)*

the time. Of the senior people, this left Ben Gascoigne, Peter Gillingham and Jack Rothwell, together with the computer group headed by Maston Beard, to work on the actual commissioning. John Pope, Roderick Willstrop and David Brown paid extended visits, and all three made essential contributions, as did Redman on two occasions. The Mitsubishi engineers stayed on until the end of April, and a Grubb Parsons team somewhat later, to see the f/8 secondary mirror installed and its tests begun.

Staffing had always been seen as a major problem. A whole new observatory had to be created from scratch, and more than anything else this was a matter of training staff in the techniques peculiar to telescope operation, and in the requirements and working habits of astronomers. The resident Siding Spring staff, mostly technicians and tradesmen, had been appointed in late 1973 and 1974. Enthusiastic and willing as they were, they clearly had a long learning process ahead. The immediate situation, however, was saved by the computer staff, comprising Graham Bothwell, Robert Dean, John Straede and Patrick Wallace. Buoyed up with what was clearly the opportunity of a lifetime, they worked incredible hours, so that at any one time there never seemed less than two of them either at the console or on the computers. They were the first night assistants, in effect if not in name. Later in 1974 a system was instituted whereby 'outside' astronomers, mostly British, came for periods of a month or two to assist generally, and at the same time to learn something about the telescope,

Figure 10.3 Commissioning. Ben Gascoigne passing a plate holder to Roderick Willstrop in the prime focus cage

and this too helped. The Board had been reluctant to appoint senior technical staff until Joe Wampler took over as Director, in September 1974. It was none too soon, because by the end of the year most of the Project Office staff, including Maston Beard, had left, but once the AAO staff began to build up the main difficulties were over.

By the time of the official opening on October 16, six months had elapsed since the primary mirror was installed. The optics were in good shape, some but not all of the drive and other mechanical problems had been overcome, the computer control programs were exceeding expectations, and the frequency of major breakdowns was beginning to decrease. The mirror was aluminised for the first time in early November, and the first really good photographic plate was obtained by Patrick Wallace on 4 December of the star cluster Kron 3 (Fig. 10.4). The non-arrival of equipment had become a problem – by December there were still no spectrographs, no photometers, no Cassegrain instrument head, no autoguider, no acquisition television and no finder telescope. The December dark run saw 'outside' astronomers invited for the first time to use the telescope, though only direct photography was then possible. Thereafter, invitations were issued most months until the telescope was made fully available in the following July.

December also saw Redman's last visit to Siding Spring. His health had been failing and he died in Cambridge on 6 March 1975. It must have comforted him to know that the telescope to which he had contributed so much was going to be as fine an instrument as he could ever have wished.

In the new year the pace quickened. The Wampler–Robinson scanner was used for the first time in March, when it created a sensation. With its completion, Wampler and Robinson were free to take a more active part in the commissioning; until then they had been almost totally occupied with the scanner and with setting up the laboratory at Epping. At the same time the staff were becoming more adept and knowledgeable, and there was no longer a shortage of night assistants. Long overdue equipment arrived, and generally a great deal was achieved. On 28 June 1975 the telescope was handed over for regular scheduled operation, a little over eight years since the decision to build it had been announced.

10.2 Some mechanical problems
With so many problems occurring in this area it would be tedious and repetitive to describe too many. We begin with the first two to be found, which were important and cannot be omitted. We have already described the single flank gear tests carried out in Japan (at a cost of close on $100,000), the detection of tooth interference, and the measures taken to alleviate it. During these tests it was also noted that as each of the 600 teeth of the main (Z) gear was engaged in turn by the drive pinion, the pressure between the two gears increased sharply, to about half a tonne. However, the full implications of this did not appear until the gears were installed and the telescope driven in right ascension. It was then found that once every tooth, or once every 2.4 minutes (the Z gear has 600 teeth), the north (lower) end of the polar axis was displaced vertically, rising and falling through about 50 microns. This might not seem a large effect – a human hair has a diameter of 60 microns – but it had important consequences, the chief of which was that it introduced a spurious rotation into the right ascension encoder, and hence a periodic wobble of an arc second or more into the sidereal drive.

The basic cause of the problem has never been identified with certainty. There were well qualified engineers, including a Project Office consultant, who expressed surprise that astronomers should expect anything better, even from a precision gear-wheel. The north journal is spherical and sits on three pads; had the design used a conventional cylindrical bearing with a thrust pad for end support the vertical movement might well have been smaller. It was established by Derek Fern, the Freeman Fox site engineer, that possibly the most important factor was that the varying pressure between the gears introduced variable distortion into the base frame. Stiffening the base frame, as described below, reduced the movement to 17 microns. When finally introduced, the right ascension damping motor also helped. Other measures were considered, but the engineers agreed that no more could be done without introducing major modifications. Instead they eliminated the symptoms.

The rise and fall of the polar axis was measured with a micro-sensor, the signal it produced was fed back via the computer into the drives, and that was effectively the end of the trouble. Although the declination drive is very similar, no analogous problem occurred there.

At about the same time it was found that the natural frequency of the telescope was 0.9 Hz instead of the expected 3.4 Hz, a bitter blow to the Project Office engineers who had put so much effort into designing it. Before long it was clear that most of that trouble too originated in the base frame, which was proving much more compliant than expected. As Rothwell put it, with feeling, *'here we have this beautiful stiff polar axis mounted on a slab of floppy jelly'*. Fortunately the Mitsubishi engineers were able to adjust the drive servo to match the new frequency without noticeable loss in performance, though the response time was now three times longer. The base frame could not be stiffened without making inadmissible structural changes to the building. Instead, at the suggestion of Derek Fern, two thrust bearing pads were added to the southern side of the horseshoe, perpendicular to its running face. They raised the frequency of the polar axis to 1.4 Hz, effectively by using it to stiffen the base frame, and if higher capacity oil pumps now had to be installed to cope with the greater oil flow, it was a small price to pay.

The polar axis now showed a tendency to oscillate sporadically at this same natural frequency with amplitudes of up to 3 arc seconds. These oscillations were not unexpected, and had already been experienced with the Palomar Telescope. They arise because the bearing friction of the horseshoe is so low that almost any disturbance can make the structure resonate. To guard against the oscillations a damping device had been designed, in the form of a damping motor and tachometer which bore against the horseshoe through a friction drive. Accelerations sensed by the tachometer caused the motor to generate a damping torque. The amplifier had not yet arrived, so Rothwell built a temporary device in which the tachometer signal was made to provide an appropriate feedback directly into the main drive. It worked so well that the original device has never been used.

While (fortunately) the Japanese were still on site, it was decided to test the emergency stop in the right ascension drive. The test succeeded beyond all expectation: it blew three power transistors and stopped the telescope for a fortnight while replacement transistors were flown out from Japan. A longer drawn-out problem occurred with the declination gearbox. From the time it was installed this produced an irregular clunking noise which could not be identified but which persuaded the engineers to reduce the slewing speed to a third. For a while one would come across worried engineers wearing headphones and listening intently to tapes, not of Hindemith or Stockhausen, but of the sound of the declination gearbox. Eventually, when the Japanese returned in 1975 for a final inspection, the gearbox was taken out and stripped down (no small undertaking), but nothing was found – no damaged tooth, no extraneous metal fragments. Back it went, clunk and all, and the slew speed was gradually edged up to normal, where it has remained without problems.

Mishaps, of course, occurred and mistakes were made. On several occasions a power failure stranded some hapless worker in the prime focus cage high above the main floor, sometimes for hours. Some sat it out until repairs were made, others opted for the perilous transfer into the cherry-picker which could be manoeuvered close alongside the telescope, lifeboat style. A potentially more serious error Gascoigne described to Redman: *'By now it was 3 am, and I had a feeling we were rushing the polar axis adjustment rather, and sure enough while doing it we forgot to disengage something, and pushed the main RA gear (the Z gear) against the encoder pinion hard enough to bend the shaft and damage a couple of teeth, not much in terms of microns but quite enough in terms of seconds of arc. Luckily Maag can regrind the damaged teeth and we emerge at the cost of not much more than red faces all round, including me.'* [2]

10.3 Testing the optics

No components of the telescope had been tested more thoroughly than the mirrors, and one might fairly ask, what else was left to do? The answer is, a great deal. First, while the primary mirror had been accepted on the basis of works tests, acceptance of the secondaries – three convex mirrors and three coudé flats – was to depend on tests carried out in the telescope. These would necessarily include tests of the cells, and especially of the mercury girdles, over which questions hung until almost the last moment. Second, although both the primary mirror and its cell had been thoroughly works tested separately, they had yet to be tested together with the tube. This had to be done over a range of altitudes down to the 70° zenith distance limit. The automatic focus had to be installed and its operation verified, likewise the effectiveness of the arrangements for ventilating the primary mirror. These led naturally to tests of dome seeing, which has proved to be a major operation in itself. Finally the operation of the cameras had to be confirmed, especially the prime focus camera with its three different corrector systems. This was important, because for better or for worse, the popular criterion for the success of a big telescope is the quality of the pictures it takes. In fact, the ability to take a sharp, well guided, long-exposure plate is as good a test as any of the two critical elements of any telescope, the optics and the drive.

Two of the earlier plates were especially significant. The first was taken in June by Roderick Willstrop, only weeks after commissioning had begun. It was a one-hour exposure of the famous globular cluster omega Centauri, taken with the triplet corrector and guided by hand. The seeing was excellent (*'Seeing is believing! RW'* reads an exultant note in the observers' book) and it was a magnificent plate, could not have appeared at a better time, and copies were widely circulated. The other was one of the first plates taken after the mirror had been aluminised. It too was hand

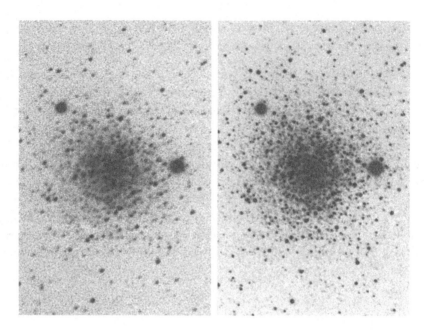

Figure 10.4   The star cluster Kron 3 taken with the ANU 74-inch telescope (left) and with
the AAT (right)

guided and used the triplet corrector, and was taken by Patrick Wallace. The object
was the Magellanic Cloud cluster, Kron 3, one of a group which had been studied
intensively by Gascoigne. He has described how he was completely transfixed by it;
it was everything and more he could ever have hoped for. The comparison between
photographs of this star cluster taken with the ANU 74-inch and the AAT shown
in Figure 10.4 might suggest why. Though he had taken a degree in astronomy at
University College London, Wallace was not a professional astronomer, and some of
the professionals around him were a little disconcerted that he should take such a
good plate at almost his first attempt. But, as we shall see, there was more of the
same to come.

Most of the testing was done by the Hartmann method, which whatever its shortcomings for workshop use remains the best way of testing a mirror in a telescope. It is
relatively insensitive to atmospheric and vibrational disturbances, which enlarge but
do not displace the spots during a time exposure. The Hartmann screen was mounted
in the tube, and the plates were taken by a number of people, with the mirror alone,
or with one of the three correctors in place. Plates were also taken over a wide range
of attitudes of the mirror. They were measured on the GALAXY machine at RGO,
and all reductions were carried out by Willstrop at Cambridge. In this and other
ways, Willstrop's contribution to the AAT optics was very considerable.

When it came to evaluating the tests, it was found that the results fell somewhat

Figure 10.5   Changing the AAT's prime focus top-end

short of those achieved in the Grubb Parsons test tower, with best figures of 80 per cent of the light within 0.4 arc seconds rather than 0.3 arc seconds. There were several reasons for this. One was disturbances introduced into the beam by warm air around the prime focus observer's cage, as the observer's body emitted enough heat to deflect the innermost rays significantly. Another was that while the Grubb Parsons results were the average over 16 plates, only two plates could normally be used for the telescope tests, hardly enough to average out any local irregularities sufficiently well. Then there was an elusive asymmetrical effect it took a long time to track down. The vital clue was provided by the knife-edge picture taken when the air was unusually steady (see Fig. 6.14 on page 94). A new feature appeared on this plate, in the form of the three parallel shadows or plumes which can be seen on the left hand side of the print, and Peter Gillingham found that they lined up exactly with three open bolt-holes in the centre section. Clearly, air inside the mounting which had warmed up during the day was being emitted in jets across the line of sight. This is a good example of the value of carrying out such careful tests, because, while the effect in itself is not great, the cumulative gain from eliminating it once and for all is well worth the trouble.

   The testing of the various secondary mirrors was straightforward but tedious. It was helped by Grubb Parsons' very accurate assembly of the tube, in the course of which it became apparent they had almost completely eliminated collimation error. In the same way, the various top-ends were found to register after interchange with a minimum of error, and virtually no hysteresis was detectable after the tube had been lowered to the horizon limit and returned to the zenith. These collimation tests owed much to a neat optical device suggested by George Sisson, one-time director of Grubb Parsons.[3] The results were confirmed most convincingly by the prime focus Hartmann tests, which are very sensitive to the sort of central coma which can arise from misalignment of the optics. Willstrop's measurements showed that coma was absent, that is, the tube was free from bending to a quite extraordinary degree. Most of the concern at this time centred around the mercury girdles, subject of so much design and redesign, trial and error. Here is another *crie de cœur*: *'Fitting the f/8 mirror, girdle and cell was a hair-raising affair, not least how I came to be present. I was measuring a Hartmann plate when Pope appeared and said "six people are assembling the f/8 and cell, not one of them belongs to the AAT, and in a few weeks they will all be thousands of miles away". The six were four from GP* [Grubb Parsons], *Pope himself and Collins the site engineer who leaves this week.'*[4] In fact, the girdles functioned very well, with only one failure in 15 years' operation.

By the end of 1974 the optics, though not completely checked, were in good shape, some highly promising plates had been taken at prime focus, and first results at the f/8 focus suggested even better, though hardly any direct photography has been done there. Observing at the Cassegrain had to be done with a temporary instrument head, little more than a steel box which contained no provision for guiding. The autofocus and autoguider still had to be commissioned, and one unexpected problem had turned up, a false image in the triplet corrector. It was unexpected because C.G. Wynne, the designer, was aware of false images and had taken steps to avoid them. However, what we had was not an image of a star, but a Fabry image of the primary mirror, produced by two internal reflections in the corrector; such a possibility had never occurred to anyone. The pseudo-image was found by Don Mathewson from Mount Stromlo Observatory, who thought he had discovered a new supernova remnant. When he saw an identical image on the next plate he became suspicious. The problem was cured by putting high-efficiency anti-reflection coatings on the two offending surfaces, at the cost of losing the ultraviolet.

The autoguider was more elaborate than some, and took up its share of computer space, but it was accurate and fast, and as well as its main function it played a useful part in developing the computer setting programs and testing the drives. Though the autoguider was on the face of it an optical instrument, its commissioning was much more a matter of computer interfacing and electronics than of optics, a situation which has become increasingly common. It was some time before the autoguider at the Cassegrain went into operation.

Photographic processing had been necessary from the beginning – the optical test programme depended on it implicitly – and one of the first tasks undertaken had been to equip a dark room. A complication then arose in the form of the new IIIaJ plates, which were just coming into use but which were too slow for most purposes unless hypersensitised by soaking in hydrogen or nitrogen. Fortunately the staff of the UK Schmidt Telescope had played a leading part in pioneering these new techniques, and their expert assistance did much to smooth the way for the AAT observers.

10.4 The computers on site

In January 1974 the computers and associated staff moved to Siding Spring. The long-deferred access to a real telescope and contact with working astronomers had an immediately stimulating effect, especially as it was not until everything was put together on site that the real way to use the computer become clear.

The computer staff consisted of Beard, Bothwell the systems engineer, Dean his assistant, and the programmers Wallace and Straede. All lived on or near the site. The control computer had first to be interfaced with the telescope and dome encoders and numerous other peripherals. It took time, and a special return visit by the Japanese (to the declination encoder) before the last problem had been solved on the last encoder. Methods of driving the telescope then had to be worked out. The end results have been described in Chapter 7, but they took a long time to perfect, depending as they did on a telescope model devised for the pointing program which had to be refined stage by stage.

The immediate operational need was for a program which would keep the dome and windscreen lined up with the telescope, specifically to within half a degree. In spite of this relatively modest requirement, the task was not trivial. The dome, shutters and windscreen had their own encoders, servo loops and drives, each a potential (and actual) source of trouble. They were awkward bulky objects, difficult to manoeuvre, and much engineering effort had to be expended before they were made to work properly. It was all finally sorted out, and the programmers could turn to the problem on which they had set their hearts, the pointing accuracy.

We must explain that in the late 1960s, one arc minute was considered reasonable pointing accuracy even for a big telescope; it was usually all that was offered and all that was asked for. It followed that star positions did not have to be known to much better than an arc minute, so that most observers chose to disregard effects like the precession of the earth's axis and refraction in the earth's atmosphere, either of which can affect a star position by at least a minute. This procedure passed muster as long as the programme stars were bright enough to be picked up in a finder, a relatively small wide-angle telescope attached to the main instrument. Meanwhile, technical developments were bringing fainter and fainter objects within reach of spectrographs and photometers, without making it any easier to find them.

Curiously, the engineers seemed more aware than the astronomers of this unsatisfactory state of affairs, and certainly were better able to cope with it. As early as

1971 John Pope had told Ben Gascoigne that the way the design was progressing, AAT users could anticipate a pointing accuracy of perhaps 3 arc seconds. Gascoigne was not unduly impressed, nor later were Redman or Wampler, all three holding that 10 arc seconds was about as much accuracy as the average astronomer would be able to use.

Accurate pointing and guiding had been stated aims of the AAT from the beginning, and were a principal argument for including a digital computer in the drive and control system, although even by 1974 it was not fully clear how the computer could best be utilised for these purposes. The first significant step towards an AAT system was taken by Bernard Pagel in October 1974, as not until then had the last problems been removed from the position encoders. Working at the prime focus with an eyepiece fitted with crosswires, he observed good telescope positions for a selection of stars chosen to cover a substantial area of sky. Then Wallace noticed that the pattern of the residuals, the differences between the observed and the tabular positions, indicated that misalignment of the polar axis was making a major contribution to the pointing errors. With appropriate and quite realistic values for this misalignment the errors were reduced to a few arc seconds. The possibility of other errors, for example, tube and horseshoe flexure, and collimation error, was immediately suggested, and a more elaborate scheme was rapidly developed.

In an early version the eight errors listed in Table 10.1 were included. Straede and Wallace drew up a list of 108 test stars, all with supposedly accurate positions, and distributed on a grid of roughly 20° squares which ran from +20° to −80° declination. On a test run they would measure the positions of the 50 or so of these stars which lay above the telescope horizon at the time. The best values of the eight unknowns were determined by a least squares fit (a mathematical technique), and in a typical run they would bring the mean error from 45″ down to 2″.5. Results at the f/8 and f/15 foci were only marginally less accurate than those at prime focus. An important point was that all the introduced quantities – horseshoe flexure is a good example – had perfectly natural physical bases. The idea of using least squares was John Straede's. In principle it could have been used at any time in the preceding hundred years, but the setting circles then used were hardly accurate enough to take advantage of it, and the calculations would have been too lengthy to justify the effort.

A further improvement came a few years later with the publication of the *Perth 70 Catalogue* of accurate star positions. It showed that some of the test star positions from the Smithsonian Astrophysical Observatory were wrong by up to several arc seconds, and with the adoption of the *Perth 70* values the AAT pointing errors were reduced to 1″.5. This extraordinary result has yet to be surpassed. As the bit size of its absolute encoders is one arc second, only limited further improvement seems possible with the AAT.

Table 10.1.

	Pointing corrections applied to the AAT		arc seconds
IH	hour angle index error	$\Delta h=$ IH	−151.8
ID	declination index error	$\Delta\delta=$ ID	−126.0
HF	horseshoe flexure	$\Delta h=$ HF sin h	−18.9
TF	tube flexure	$\Delta z=$ TF tan z	+2.0
NP	ha/dec axis not perpendicular	$\Delta h=$ NP tan δ	−1.6
CH	collimation error	$\Delta h=$ CH sec δ	+10.0
MA	polar axis azimuth error	$\Delta h=-$MA cos h tan δ	+17.6
		$\Delta\delta=$ MA sin h	
ME	polar axis elevation error	$\Delta h=$ ME sin h tan δ	+79.4
		$\Delta\delta=$ ME cos h	

Δh and $\Delta\delta$ are the errors in hour angle and declination produced by each of the individual maladjustments. The right-hand column shows typical errors in arc seconds. ME is relatively large because the instrumental pole is located near the refracted pole, not the true pole.

Though immediate credit for this result must go to the programmers, it could hardly have been achieved without the sound structural design of the mounting, and the accuracy and the freedom from hysteresis of the gears and the optical support systems. Note also that it was not necessary to introduce corrections for gear errors into the computer program. As explained earlier, the effects of the rise and fall of the north end of the polar axis were eliminated by measuring its displacement with a transducer and feeding the information into the computer. The computer could then make the appropriate adjustments to the parameters IH and ME in Table 10.1.

G.H. Hardy, a celebrated mathematician with strong convictions, once wrote a book in which he urged a more rigorous approach to English mathematical analysis; it was said that in his book he sounded like a missionary talking to cannibals.[5] Here is Patrick Wallace addressing the astronomers:

> The pointing of the AAT is so good that it is irresponsible to begin observing with positions to the traditional one arc minute when more accurate figures could have been obtained; time is sure to be wasted. With positions to the nearest arc second correct identification is almost automatic

and observing can be under way within seconds of the end of the slew. Poor positions, on the other hand, can lead and have led to enormous wastage of time; twenty minutes is not uncommon and over an hour has been reported.

Note that certain procedures widely regarded as good enough are simply not so for the AAT. Proper motions, for example, are almost always neglected yet are often significant when arc seconds matter. Also the practice of entering the mean place for the start of the current year for an apparent place is quite wrong, as is entering an apparent place for, say, June 1975 as a mean place for 1975.5.[6]

In the AAT control system tracking and pointing are treated as one problem, with any inadequacy in the pointing model reflected as a long-term tracking error. As would be expected, long-term tracking is very good. Tracking is discussed later in the chapter, in conjunction with the autoguider.

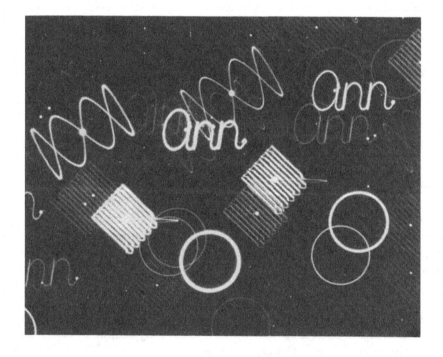

Figure 10.6   The Ann plate. This was taken to demonstrate the variety of patterns which the computer can cause the AAT to trace on the sky. Stars which happen to be in the field mark out the patterns made by the moving telescope. The lines in the raster scans are separated by 10 arc seconds.

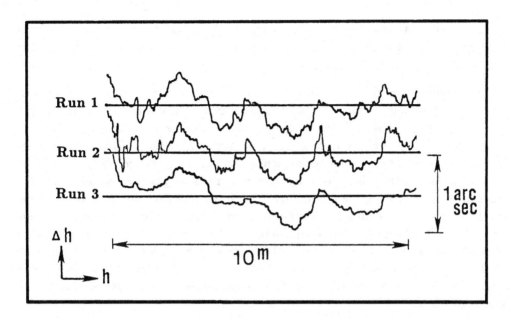

Figure 10.7   The autoguider error signals produced by three successive runs over the same 10-minute range of hour angle, made on a night of very good seeing. This range includes three teeth on the main (Z) gear. Note that the linear scale around the circumference of this gear is 9 microns per arc second.

From the pointing program Straede and Wallace went on to ventures which were inconceivable before the time of the AAT. They have enabled the telescope to carry out spiral scans, raster scans in any orientation, and indeed any pattern which could be programmed. The scan programs have been used to great effect by the infrared astronomers, for finding sources with only approximate positions, and for mapping small areas with a pencil beam. A selection of these programs in the form of a number of superimposed exposures is shown in Figure 10.6. The recipient of this unique tribute was Ann Savage, then on the staff of the UK Schmidt Telescope. The corresponding slide was first shown at a conference at the Massachusetts Institute of Technology attended by Wallace and Bothwell. Wallace's talk was the first public account of the AAT control program. He said to Bothwell afterwards *'it didn't seem to go down too well, no response at all'*. To which Bothwell replied: *'Not so. You left them dumbfounded, speechless – and they're still in a state of shock.'* [7]

The commissioning of the autoguider enabled the pointing calibration program to be made completely automatic. The autoguider was mounted on axis, and once it had acquired a star it would centre the telescope to a tenth of an arc second within a few seconds. The position was marked, the solution for the constants updated, and simultaneously the telescope moved onto the next star. It was uncanny to watch the telescope pursuing its solitary, erratic way to and fro across the sky, like an ungainly pavan performed against the rumblings from the dome and the more subdued drone of the gearboxes. It takes less than an hour to complete a program which normally includes about 50 calibrating stars.

The autoguider could be used to make quantitative examinations of the drive performance by plotting the right ascension error signal against time, as in Figure 10.7. The cumulative error signal could also be plotted. The figure shows successive runs made across the same range in hour angle in very good seeing. The range included three main gear teeth, and tooth-frequency and other errors are obvious. Bear in mind that the linear scale at this wheel is nine microns per arc second. This is a powerful method of studying the quality of both RA and declination drives, and has revealed various slow drifts, long and short period cyclic errors, and sundry irregularities. Despite these, it is found that blind tracking errors are generally within ± 1 arc second for an hour or more, and natural frequency oscillations are less than 0.1 arc second peak to peak.[8] In declination, only small, rapid variations are seen, with amplitudes less than 0.1 arc seconds. Because the declination speeds are so slow, this may be due to stiction.

The beam-switching program should be described because it is used so frequently and has been so successful; it is closely associated with the sky subtraction referred to in Chapter 5. Many modern instruments either alternate observations between object + sky and sky, or have dual (or multiple) channels which enable them to observe object + sky and sky simultaneously. In the latter case the object is often alternated between the dual channels. To set up the scheme a star is first centred in channel A, which may be a mark on the video screen, and the observer presses button A on the console. A star is then centred in channel B, and button B is pressed. It can then be interchanged between channels in a very few seconds, no matter (within reason) how far apart they are. Moreover, this relation will hold regardless of the position angle to which the instrument may subsequently have been rotated. This is a good example of the way in which computer control can make observing easier, quicker and more certain.

For a final example we turn to a program for which John Norris (MSO) required spectra of a sizeable number of red giant stars in the globular cluster omega Centauri. Accurate positions for many of the cluster stars having previously been measured at RGO, Norris was able to load those he wanted into the computer. At the end of each exposure he merely had to tell the computer 'next star' and that star would be on the spectrograph slit within seconds. Even in a field as difficult as that of omega Centauri, identifications were never in doubt. The only dead time was that taken to transfer the data from the last exposure to a magnetic tape. This efficiency which the computer brings to observing, the absence of errors, and the way it frees the observer from the repetitive tedium of the process, confer great advantages.

10.5 Instruments and accessories

The first steps towards the auxiliary instrument programme were taken at the third meeting of the JPC, in March 1969, when it was decided to make provision for the following instruments. The total cost was estimated at about a million dollars.

1. direct photography cameras at the prime and f/8 foci (Project Office)

2. low dispersion (fast) Cassegrain spectrograph (MSO, Alex Rodgers)

3. intermediate dispersion spectrograph (RGO, Don Palmer)

4. photoelectric photometers (MSO, with UK contribution)

5. high dispersion coudé spectrograph (Herman Wehner, Roger Griffin)

At that time, with the concept of 'common user' or 'people's instruments' very much to the fore, it was intended that all the above would be common user instruments, in the sense that they would be simple enough to be within the capacity of the average astronomer, as opposed to more specialised instruments like image intensifiers and electronic cameras. Alex Rodgers (MSO) and Vincent Reddish (RGO) took the lead in formulating these plans, about which discussion ranged widely; as was to be expected, astronomers were much more interested in the instruments than in the telescope itself. Difficult decisions were called for, as there was every prospect of major developments in the field before long, especially the production of detectors which would combine the high quantum efficiency of the photoelectric cathode with the extended-field capacity of the photographic plate. Such a development did of course eventuate, in the form of the Robinson–Wampler image dissector scanner (IDS), first used on the AAT early in 1975. Its impact was revolutionary, and it overshadowed the instruments of the original suite to such an extent that the only one to have survived is the RGO or intermediate dispersion spectrograph.

If the AAO has a workhorse it is the RGO spectrograph, which having been in constant use since mid-1976 merits a brief description. It is used at the f/8 focus, and is a massive instrument with a 15-cm collimator and a choice of three gratings and two cameras, the latter with focal lengths of 25 and 82 cm. Dispersions from 5 to 130Å per mm are available in first or second order. The usual detector is an image photon counting system (IPCS) of the type developed by Boksenberg when he was at University College London, and which came to supersede the IDS. Like almost all AAT instruments, the RGO spectrograph is arranged for complete remote operation, with guiding, when necessary, carried out with the help of a sensitive slit-viewing television camera. Some years later its range was significantly extended when the faint object red spectrograph (FORS) was built. This is a very efficient low-dispersion spectrograph which uses the mounting and slit-head of the RGO spectrograph but none of the RGO optics. Its own optics, in which it has been possible to omit the collimator, were designed by C.G. Wynne.

Of the other instruments, only the dual-beam photometer was used to any extent, and within a few years it had been replaced by electronic 'area' detectors. It had been designed from the outset for remote programmable operation, and was the first instrument to be driven from the instrument computer. It had its night of glory when, supplemented by a new high-speed data-processing system, it was used to identify the optical counterpart of the Vela pulsar. This was noteworthy among other reasons as the faintest object to have been detected up to that time by an optical telescope. The fast spectrograph ran into a series of optical problems associated with the cameras and was never seriously used, while the coudé spectrograph, though designed in some detail, was postponed and then cancelled, partly because demand for coudé spectroscopy had fallen off, but mostly because no funds were available for it in the budget.

The IDS was a copy of an instrument Robinson and Wampler had built for the 120-inch Lick telescope. Its key element was a so-called dissector tube in which the signal was recorded not on a photographic plate, but picked off TV-style by a scanning electron beam and stored in a computer memory. At any time the observer could see how he was progressing by displaying the signal on a video screen, and this, coupled with the linearity that enabled the sky background to be subtracted accurately, conferred enormous advantages. The instrument was operated from a computer keyboard (the computer was a surprisingly modest PDP-8 shown in Figure 10.8); no particular manual or instrumental skill was required, and the 'common user' concept had vanished overnight. The power and versatility of the control system were astonishing, and it was this more than anything else that pointed the way to the future. When asked, as he often was, how much the IDS cost, Wampler used to reply: *'the real cost was the ten man-years it took to write the control software'*. Perhaps Don Mathewson summed it up best, when after his first run on the IDS he returned to the Lodge at daybreak *'with more data from one night than I'd have got in a year on the 74-inch'*.

The joint success of the telescope and the IDS had another, unforseen consequence. As Hoyle has remarked, the ANU did not take kindly to its defeat over the AAT management issue: *'Having lost the struggle it was time for Olin Eggen to retreat . . . with a rueful grin, and nobody then would have thought any more of it.'* [10] As it was, the whole commissioning period was marred by a series of petty squabbles and restrictions, over the Lodge, the workshop (more than half the cost of which the AAT Board had met), parking space, and, it seemed, every other aspect of mountain-top life. But suddenly all was sweetness and light. The past was forgotten, the old problems were swept away on a wave of euphoria, and all that the Mount Stromlo astronomers now looked for was time on the new telescope. And time they got in

Figure 10.8 Joe Wampler, the first Director, seated with the computer and associated hardware which controlled the Image Dissector Scanner. The PDP-8 computer occupied a relatively small part of the rack.

good measure, about half the Australian share in fact, which at some 80 nights a year was much more than any other institute in either country. Further, the AAO was no longer perceived as a competitor for funds or staff, but rather as a major accession to the overall strength of Australian astronomy.

Various auxiliaries call for comment. The television system was delivered by its American manufacturer, Quantex, in February 1975. Once installed, and following a good deal of adaption, it was found that with an integration time of a few seconds it could detect a twenty-second magnitude star. This very satisfactory performance was the death-knell to earlier plans of finding faint objects by off-setting, and took much of the pressure off the commissioning of the Cassegrain acquisition and guidance head. In the following year its performance was enhanced further by a digital memory built by the AAO electronics staff, and this confirmed the status of the television system as a permanent part of the telescope.

The acquisition and guidance head into which so much design and other effort had gone was delivered at about the same time as the television system. It is installed semi-permanently at the Cassegrain focus, and is the interface for all Cassegrain instruments. Probably its most often used property is that it can be rotated, and thereby allows a spectrograph slit to be set at any orientation to a field. It was some time before the probes and autoguider were made to work – both are controlled

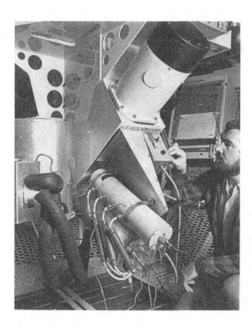

Figure 10.9   John Barton, electronics engineer, at the Cassegrain focus with the Image
Dissector Scanner. It is mounted on the Boller and Chivens spectrograph.

from the console – but these too have found a niche, in the form of guiding the long
exposures sometimes called for by multi-object spectrograph plates.

In a sense the commissioning period was irrelevant to the instrument program.
Aided by a generous funding policy and a strongly supportive Board, the program
has continued at a vigorous pace ever since commissioning, as was necessary if the
AAT were to maintain its position as a leader in the field. Throughout this period
the instrument computers have played a steadily expanding role; one can hardly now
conceive of a new instrument that is not fully interfaced with and controlled from a
computer. On the other hand, once the programs for operating the telescope had been
fully developed the control computer was left to itself, except for maintenance and
the occasional upgrading of components like disc drives. Contrary to fears expressed
at one time to the JPC, its reliability has become a byword, and 17 years since its
purchase there is still no move to replace it. The telescope itself has long since been
taken for granted, and as observers now rarely see it, certainly not at night, when
it is kept in total darkness, its status relative to the instrumentation has diminished
almost to that of a computer peripheral.

Looking back on the commissioning, the dominant impression it leaves is how
rapidly the style of observing changed over that period of not much more than a year.
Observers went to SSO expecting to work much as they had always done, guiding
by eye the image of the faintest of galaxies on an all but invisible spectrograph slit,

Figure 10.10   The 3.9-metre Anglo-Australian Telescope

identifying a critical star by matching a crowded field seen in an eyepiece with that on a dimly lit hand-held chart, changing plateholders and making all instrumental adjustments by hand, in total darkness. Certainly more accurate setting, better guiding, and enhanced television viewing held promise of easier observing. But they were in no way prepared for what actually awaited them – a warm, well lit control room, where the observations unfold on video screens, and the telescope is run by the resident technician because he knows the control program so much better than you, the observer, ever will.[11] Not only the old-time observers' skills but the observers themselves are steadily becoming redundant. On the other hand, the output of data is prodigious – the yield from one good night can set a research team up for months, and more than ever the real work of observing has come to centre upon a computer terminal back at home base, and the massive reduction programs which are so central a feature of present-day astronomy.

Notes to Chapter 10

1 Redman R.O., report on his visit to Siding Spring, March 1974
2 Letter Gascoigne to Redman, 29 May 1974
3 This device is shown in Figure 10.11. Four small flats were cemented inside the central hole of the primary mirror, equally spaced around it and made parallel to the optical axis. Their parallelism can be verified only from the centre of curvature, and must be done in the manufacturer's test tower because the centre, 25.4 metres from the mirror, is otherwise inaccessible. But once done it allows an alignment telescope

mounted behind the mirror to be pointed quite accurately along its optical axis, and from then on the operation is simple.

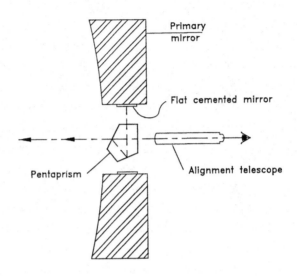

Figure 10.11   An instrument for checking collimation devised by G.M. Sisson and used with the AAT

4 Letter from Gascoigne to Redman
5 This comment was made by Hardy's long-time collaborator J.E. Littlewood, and is quoted in the preface of the book by Hardy G.H. (1937) *A Course in Pure Mathematics*, 7th ed. Cambridge University Press.
6 Observer's Guide to the AAT, 1976 ed. §3.3.1.1
7 Wallace P.T. (1975). Proc. MIT Conference on Telescope Automation, p.284. See also Wallace (1978). ESO Conference on Optical Telescopes of the Future, F. Pacini *et al.* (ed.) pp.123–131
8 Gillingham P.R. (1984). Proc. IAU Colloquium No 79, M-H. Ulrich and K. Kjar (ed.) p.415
9 Wallace P.T. (1979). 'Telescope Pointing Investigations at the Anglo-Australian Observatory' (internal report)
10 Hoyle F. (1982). *The Anglo-Australian Telescope*, University College Cardiff Press, p.17
11 John Whelan (1976) wrote a lively, first-hand account of the AAT as it was in Wampler's time. *Qrt. J. R. Astr. Soc.* **17**, 306

11   The UK Schmidt Telescope

11.1 The proposal

The work of the 1.2-metre Schmidt Telescope on Siding Spring Mountain has played such a large part in the success of the AAT that an account of the origin and development of the telescope must be included in any comprehensive history of its larger companion. We shall describe later in the chapter how well the telescopes complement each other. Here it suffices to say that the particular attribute of the AAT is the ability to collect and detect light from very faint or distant sources from a small area of sky, while the Schmidt Telescope has a wider field of view and is, in effect, a very large and accurate camera. A major use of the Schmidt is to make photographic atlases of the sky and to reveal objects which may merit more detailed examination by the AAT and its comprehensive range of instruments.

When the British and Australian Governments reached agreement in 1967 to proceed with the construction of the AAT, optical astronomy in the northern hemisphere was dominated by the large American telescopes, particularly the 200-inch on Palomar Mountain. A companion 48-inch Schmidt telescope had mapped the northern sky in unprecedented detail in the early 1950s. In sharp contrast, astronomy in the southern hemisphere was dependent largely on scattered, independently run telescopes mainly in Australia and South Africa, the largest of which in each country was 74 inches in aperture. In South Africa, the British community operated two observatories, the Royal Observatory, Cape of Good Hope, in Cape Town, and the Radcliffe Observatory in Pretoria which had been transferred there from Oxford in 1935. However, there was no southern hemisphere sky atlas to match that produced by the Palomar Schmidt telescope. In retrospect, it is surprising that there were no plans for the AAT to have an associated Schmidt telescope from the beginning, but no doubt 'one step at a time' was the policy considered most likely to succeed, and the AAT Agreement referred to only one telescope.

The first proposal for a UK Schmidt Telescope appeared in the report of the Southern Hemisphere Review Committee. This had been set up by the Science Research Council (SRC) in December 1967 to examine the facilities for British astronomers in the southern hemisphere. In particular, the impending cost of the contribution to the AAT prompted a quest for economy in British expenditure on astronomy in South Africa, while at the same time the South African experience could perhaps offer some lessons on how best the AAT could be operated from the British point of view. The Committee was chaired by Sir Fred Hoyle, and included J.F. Hosie, R. Wilson and R.O. Redman. One important result for British astronomers was the recommendation which led to co-operation between the South African Council for

Scientific and Industrial Research (CSIR) and the SRC through the formation of the South African Astronomical Observatory (SAAO).

The Cape and Radcliffe Observatories were acquired by CSIR and became the SAAO. At the same time their main telescopes, especially the Radcliffe 74-inch and the Cape 40-inch, were transferred to a new site at Sutherland, in the very good climate of the Great Kâroo. The Cape Observatory buildings at Cape Town became the headquarters of the SAAO. An annual contribution from the SRC to the SAAO ensured that British astronomers continued to have access to the South African facilities. Both the Cape Observatory, which was founded in 1820, and the Radcliffe Observatory had distinguished records of achievement, and their amalgamation marked the end of an era in southern hemisphere astronomy.

However, it was the second and unexpected recommendation from the Review Committee which led directly to the Schmidt Telescope at Siding Spring Observatory.

It had always been clear to the members of the Review Committee that British users of the AAT would face special problems in using the telescope so far from their home base. Their examination of the problems experienced by users of the telescopes in South Africa, especially the lone instrument of the Radcliffe Observatory, reinforced this feeling. At this time, the arrangements for staffing the AAT and the siting of a scientific base, should there be one, were undecided. It was not known how operations of the telescope, its management and the development of its instrumentation would be organised, or how many resident astronomers there would be. The Committee was doubtful whether the AAT could be used efficiently by visiting British astronomers unless there were a core of resident astronomers having good scientific and administrative links with the British astronomical community. This was not to deny the vital role of visitors bringing inspiration and fresh ideas to the operation, but simply to recognise that short-term visitors, however talented, could not provide the continuity of experience needed to maintain a complex facility in first class condition. The Committee concluded that some at least of these difficulties could be avoided if there were a small British base at Siding Spring adjacent to the AAT. This would be a scientific centre for visitors from Britain and a point of contact between the AAT and half its potential users.

It was the linking of these ideas, initially by Hosie and Wilson, with the recognition that the progress of astronomy with the AAT would be greatly enhanced if there were an associated Schmidt telescope at Siding Spring capable of mapping the southern sky, that led to the Review Committee's recommendation:

> Having considered the long-term British requirements of astronomy in Australia the Committee is convinced of the need to establish a British supporting unit alongside the Anglo-Australian Telescope at Siding Spring

with offices at Canberra on or near the campus of the Australian National University (ANU). They believe that this is essential for the full and efficient use of the UK's share of time on the 150-inch telescope. Such a unity would provide a base for long and short-term users of the AAT from the UK and also, if desired, continuity for fundamental programmes of research. In reaching this conclusion the Committee particularly noted that the real strength of Palomar Mountain does not come from the 200-inch telescope alone but also from the powerful array of supporting instruments – a 48-inch Schmidt telescope and a 60-inch on Palomar itself, backed by a 100-inch and another 60-inch on Mount Wilson. While such a rich grouping of instruments can hardly be contemplated at Siding Spring, the Committee considers it essential, if Siding Spring is to achieve its full potential, for the UK to provide supplementary observing facilities in an enclave. It would clearly be sensible that these should be complementary to those provided by the ANU at Siding Spring and that agreement for mutual access by Australian and British astronomers to all telescopes should be sought in order that for any programme the most appropriate instrument should be used. At present ANU has three telescopes there – of aperture 16, 24 and 40 inches. There is also a 74-inch at Mount Stromlo. The Committee believes that . . . the UK should consider first the provision of a 48-inch Palomar-type Schmidt instrument, if possible within the next five years, and then give second priority to an optical telescope of advanced design and order of aperture 60 inches . . .

In a now famous letter in correspondence to *Nature* in 1972[1] Geoffrey Burbidge criticised this decision when he listed key mistakes in British scientific policy. One mistake, he wrote, was

The decision to build a large Schmidt telescope in Australia without appearing to know or care that it is being put on an inferior site and that the ESO group will have a similar telescope operating in Chile this year.

Burbidge was not alone in holding these views. Not only did ESO have a long start on the British, but of the 23 sizeable Schmidt telescopes in existence at that time, 15 were located in Europe, and it was widely held that if there was one field in which the Europeans were expert it was in the operation of Schmidt telescopes.[2] Nevertheless by common consent, the best Schmidt was the Palomar 48-inch telescope, which had produced the famous survey of the northern sky. A British southern survey would invite direct comparison, and the Palomar standard was going to be hard to meet. That the British survey met the challenge by a handsome margin is now history.

Fulfilment of the Review Committee's proposals required progress on two fronts.

Figure 11.1   View from the Observatory to the north with the Schmidt Telescope building

The Australian authorities, in particular the University, the owner of Siding Spring Observatory, had to be willing to accommodate the proposed Schmidt telescope with its associated access road and requirements for services on the mountain. They had to be assured that the necessary extra housing would not interfere by increased lights or in any other way with the established telescopes, and that satisfactory financial arrangements could be made. In Britain, the various astronomy committees of the SRC, the Council itself, the Department of Education and Science and ultimately the Treasury, had to be convinced in various ways that the proposal had sufficient merit to justify the expenditure at a time when funds allocated to astronomy were under severe pressure. The initial capital cost estimate was £650,000, later increased to £750,000, spread over the financial years 1970–71 to 1973–74, with estimated annual running costs of £45,000.

The Southern Hemisphere Review Committee had visited South Africa in April and May 1968, and its report was ready for consideration in Britain in late May. The proposal for the Schmidt telescope was not destined to have an easy or speedy passage. The first reaction of the Astronomy Policy and Grants Committee (APGC) was favourable to the general ideas put forward by the Review Committee regarding the Schmidt. It supported plans to reduce British expenditure on astronomy at the South African Observatories by striking a deal with CSIR. However, it was soon realised that the resulting economies would go only a small way towards meeting the cost of the Schmidt telescope. The Astronomy Space and Radio (ASR) Board of the SRC, the parent body of the APGC, considered the matter for the first time in December 1968 and also reacted favourably in general terms, although at this stage the precise financial implications remained to be clarified. In the following months the case for the Schmidt telescope was debated widely, particularly by the leading radio astronomers in Britain. Sir Martin Ryle was unenthusiastic about additional southern hemisphere facilities at this time, while Sir Bernard Lovell viewed the financial demands as a serious threat to a proposed major new radio telescope, the Mark V, for Jodrell Bank. The essence of the problem was that the likely financial allocations to the APGC over the years in question, 1970–71 to 1973–74, were very largely committed to projects already under way. The only flexibility appeared to be related to the proposed Schmidt telescope and the proposed Mark V radio telescope. The timing of the incidence of expenditure would be crucial. It was, and still is, a feature of the British method of controlling public expenditure that money allocated for spending in a particular financial year is forfeited if not actually spent in that year. Although the APGC wished to give priority to the Mark V radio telescope, it was argued by Hosie and eventually accepted that in view of the many preliminary steps still inescapable for such a large project, any significant funds allocated to the radio telescope in the following few years simply could not be spent in those years. They could therefore be diverted elsewhere – to the proposed Schmidt telescope for instance. Unpalatable though this was to the radio astronomers, it offered the only reasonable alternative to the funds being diverted out of astronomy altogether.

11.2 The project

By June 1969, Woolley, Hosie and Hoyle, the British members of the Joint Policy Committee for the AAT, had obtained Australian agreement to the proposed Schmidt at Siding Spring, and had earmarked a site east of the AAT. This understanding was given formal shape by a document in October 1969, *A Basis of Co-operation between the ANU and the SRC of the UK in the building and operation of a 48-inch Schmidt Camera Telescope on Siding Spring Mountain.* In addition, Horace Babcock, Director of the Hale Observatories, had agreed to make available a complete set of drawings of the Schmidt telescope on Palomar Mountain. In recognition of his generosity, the SRC agreed that Palomar astronomers should be offered some access annually to the Schmidt at Siding Spring. The project was accepted by the APGC and the ASR Board in late 1969, and formally approved by the Council in January 1970. However, the necessary Government approval was not obtained until early 1971, a delay caused by a change in the British Government after the 1970 election. The project was finally under way two and half years after the initial recommendation.

This unexpected delay in obtaining authority to proceed made it more than ever important to go ahead with all speed. The progress with the construction of the AAT and the probability that it would be operating in 1974 gave added impetus to the Schmidt project. The key to rapid progress would lie with management of the project. Here a very wise choice was made in the appointment of V.C. Reddish of the Royal Observatory, Edinburgh, to head the whole operation. At Reddish's insistence, an independent Schmidt Telescope Project was set up based in Edinburgh but completely independent administratively and financially, and run directly by the SRC under the guidance of an advisory committee comprising astronomers who were the potential users of the telescope. Later when the telescope was completed, the Project became the UK Schmidt Telescope Unit, and retained its independent identity within the SRC until Reddish became Director of the ROE and Astronomer Royal for Scotland in October 1975, when it was merged with the Observatory.

Reddish had shown his skill in practical matters by the way in which he had taken up and successfully completed a complex astronomical plate measuring machine known as GALAXY. He was also well acquainted with astronomy in Australia, having been a consultant on various technical aspects of the AAT itself, and was well positioned to tackle the new Schmidt Telescope. Reddish's brief was to design, build, commission and operate the Schmidt Telescope. The terms of reference from the SRC required him to copy the Palomar Schmidt design with a minimum of modifications. The need for speedy progress and the backing of the Astronomer Royal for the Palomar design both apparently favoured a close copy, subject only to modifications essential as a consequence of the changed site.

Reddish wisely began with a close examination of the Palomar telescope. He used it for two nights himself, and was able to discuss its performance freely with the co-operative Palomar staff. He soon realised that radical design changes would be needed if the new Schmidt were to be fully up-to-date and consistent with the new image being given to British astronomical facilities. It is not appropriate here to detail all the technical features with which Reddish had to grapple; a few examples will show the nature of the problems. The intended use of the UK Schmidt required that the plate images be of the highest possible quality to take advantages of the automatic measuring machines such as COSMOS, the successor to GALAXY. This in turn demanded redesigning the drive and anti-backlash systems, the plateholder and focussing systems, and the support system for the Cervit mirror. It also required replacing the visual guide telescope with a fully automatic acquisition and guidance system. The possibility of using new photographic emulsions which had much better resolution than those used for the Palomar northern survey also encouraged the adoption of tighter optical and mechanical tolerances. By making the whole telescope more rigid it would be possible to mount thin objective prisms at the entrance aperture, and thus open up a whole new field of slitless spectroscopy of faint stars and galaxies.

In the light of these developments and to take full advantage of them, so much redesigning appeared necessary that a fresh start was far more likely to succeed than a hybrid Palomar-ROE design. This led Reddish to issue a detailed performance specification which would inevitably result in the design features needed in the new telescope, while staying within the formal contractual constraints of the SRC. Tender action resulted in Sir Howard Grubb, Parsons and Company of Newcastle–upon–Tyne winning the contract for the telescope. Manufacturers in Germany and Japan claimed that Grubb Parsons could not build the telescope in under three years. Nevertheless, the British company designed and built it meeting the tight specification in all respects. Moreover, it achieved this in two years, which was a remarkable achievement. If the telescope had taken longer to complete, there was a real fear that the funds allocated by the SRC would have been forfeited. Reddish was responsible also for the design of the building and for all ancillary equipment, giving him a clear overview of the complete system needed for the eventual production of sky survey photographs.

The particular feature of Bernhard Schmidt's original 1931 design of telescope which makes it so important for sky mapping photography is the wide field of view compared with that of conventional large telescopes. This field is 6.4 degrees square for the UK Schmidt compared with, at most, one degree for the AAT. Photographs are taken on glass plates 356 mm square and one mm thick. The plates have to be thin so they can be bent to match the curved focal surface of the telescope. Special photographic emulsions have been developed for the very low light levels and very long exposure times, typically about one hour, used in astronomy, with sensitivity to different spectral regions of interest. The quality of materials for astronomical photography had been dramatically increased since the time of the Palomar survey, with the introduction in the mid-1960s of Eastman Kodak's IIIaJ emulsion. The fine grain, high contrast and high resolution of this material were properties ideally suited to recording the small-scale, highly detailed images produced by a Schmidt telescope. The emulsion was also designed to record the faintest signals buried in the ever present airglow of the night sky. Although these properties were highly desirable, the emulsion itself was very slow and it was by no means certain that the hypersensitising techniques then in use would increase the speed to a useful level. Indeed, some astronomers at the time criticised Reddish's decision to use the new emulsion. Although it could yield excellent results for specific objects, they argued that its use was impractical for a uniform large-scale sky survey. However, the Schmidt group at ROE persisted and developed a satisfactory technique for hypersensitising IIIaJ plates by prolonged soaking in nitrogen gas for many weeks. Later, in 1974, the Eastman Kodak company suggested that soaking in hydrogen gas for a few hours could yield greater gains in speed. The Schmidt Telescope Unit successfully pioneered the routine use of a combination of the two gas treatments for both the blue–green sensitive IIIaJ and the later red sensitive IIIaF emulsions. Also, the meticulous photographic techniques adopted by the Schmidt Unit staff have set new standards for Schmidt observers everywhere.

11.3 The optical system

The optical system of the telescope is shown in Figure 11.2. The largest component is a spherical mirror of 1.8 metres diameter and 3.06 metres focal length. Used alone

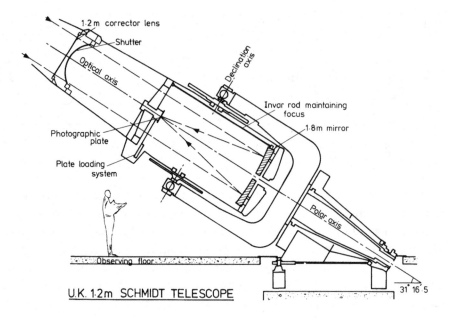

Figure 11.2   The optical system of a Schmidt telescope

this mirror would produce blurred images of unacceptable quality, but they can be
made point-like by a specially contoured, almost flat, plate of 1.2 metres diameter,
located at the centre of curvature of the mirror. The maximum departure of the
plate from flatness is about two microns. With this correction, and within a limited
wavelength range, the system is capable of producing virtually perfect images over
the whole area of the photographic plate. However, because the refractive properties
of glass change with wavelength, the images at the limit of the photographic range,
say at 330 nm and 1200 nm, will exceed four arc seconds diameter, which again
are unacceptable. This restriction can be removed by replacing the simple correcting
plate with an achromatic plate made of two glasses with different refractive properties,
shown in Figure 11.3. Such a plate was designed for the UK Schmidt by C.G. Wynne
and made at Grubb Parsons under the direction of David Brown. It reduced the
image diameters at the ends of the photographic spectrum to less than a third of
their previous values (a tenth of the area), and the optimum performance was now
gained over the whole of the available spectrum. This plate was the first of its type
to have been made, and was a considerable technical achievement. This was not only
because of its great size, and because the components had to be figured respectively
three and four times more deeply than the original plate, but because the process of
cementing them together with the total exclusion of all bubbles was extraordinarily
difficult. It has been highly successful, and Grubb Parsons has since made similar
achromatic correctors for the ESO 1-metre Schmidt in Chile and the original Palomar
48-inch Schmidt telescope.

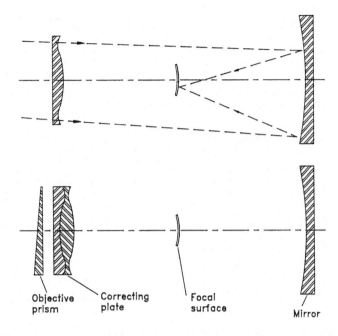

Objective prism   Correcting plate   Focal surface   Mirror

Figure 11.3   Upper diagram: the optics of the Schmidt Telescope as originally received. Lower diagram: the achromatic correcting plate and objective prism. Either of the prisms (there are two) may be inserted or removed at will. The asphericities of the correcting plates are greatly exaggerated; that on the inside surface of the lower plate, separating the two different glasses, is about three times that of the upper plate.

In May 1973, the telescope was delivered to Siding Spring; the first plate was taken with the guidance system operating in July 1973, and the telescope was formally opened by Professor Bengt Strömgren, President of the International Astronomical Union, on 17 August 1973. The crucial date for formal acceptance by the SRC was 3 September 1973, just three days late in terms of the original contract.

Many technical improvements have been made since that time; we have already referred to the achromatic corrector lens and there have been further advances in the hypersensitisation techniques for various emulsions, especially at infrared wavelengths. However, the enhancement probably of greatest relevance to the AAT was the introduction of objective prisms, made possible by the foresight of Reddish, as we have mentioned above. The basic optical system of the telescope is designed to produce as near as possible point images of stars. When a large glass prism is interposed at the top end of the telescope, each image is spread into a spectrum. The proportions of such prisms are unusual; they are slightly wedge-shaped discs 1.2 metres in diameter, and varying in thickness by only a few tens of millimetres across their width. Two such prisms are available for the telescope: one of low dispersion

Direct plate   Prism plate

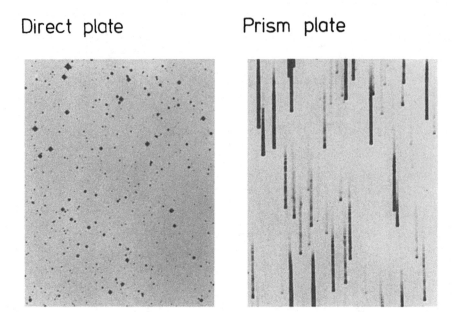

Figure 11.4   These two photographs of the same area of sky show the comparison between
a normal Schmidt photograph and one taken with an objective prism in position. On the
right each image is dispersed into a small spectrum. *(Photo: ROE)*

with an apex angle of 44 arc minutes, and another of intermediate dispersion with an
apex angle of 2°11'. The prisms can be used individually, or together in parallel or
anti-parallel positions, allowing a choice of dispersions. Figure 11.4 is an example of
the same area of sky photographed with and without the objective prism in position.
Without the prism the stars are almost-point images; with the prism each image is
dispersed into a small spectrum. The objective prism technique gives a very powerful
method of identifying some of the images on the Schmidt plates. In particular, the
spectra of quasars at high redshift often show strong emission lines and absorption
troughs, and to the experienced eye are readily distinguished from the spectra of
stars. Interesting quasars and other objects identified in this way can be examined
in greater detail with the AAT, and the two telescopes have a long and successful
record of discovering a succession of quasars, the most distant known objects in the
Universe. Like the achromatic corrector, these large prisms are unique in the world,
and owe their success to the superb skill of David Brown and his colleagues at Grubb
Parsons.

In the first year there were many problems to sort out, and Reddish believed a
team of technical experts could perform the survey following closely specified procedures. But in 1974 the Deputy Project Officer from the Schmidt Telescope Unit
in Edinburgh, who was a research astronomer, came to Siding Spring to get the sky
survey under way. He solved the awkward problems associated with the hydrogen

sensitising of the IIIaJ plates, raised the photographic technology, refined the mechanical and optical adjustment at the telescope and its autoguiding equipment, started the survey, and generally established the high standard of the work which the telescope continues to produce. He saw that the policy of having research astronomers involved, especially in charge of the operation at Siding Spring, was maintained. Some 12 years later the same person, Russell Cannon, would become Director of the Anglo-Australian Observatory.

11.4 The scientific work

Once the telescope had been commissioned fully, its observing programme developed along two lines. First, there were the surveys of the southern sky – indeed the most important reason for building the telescope – which have been carried out in conjunction with the 1-metre ESO Schmidt telescope in Chile. Secondly, there have been large numbers of special sky photographs taken for individual research projects of astronomers in Britain and elsewhere. The first Southern Sky Survey, covering the sky for declinations south of $-17°$, required 606 plates, with plate centres $5°$ apart. The UK Schmidt Telescope Unit took plates in the blue wavelength region while ESO took red plates. This survey is now complete and other surveys are in progress, one particularly in the near infrared, and another extending the Southern Sky Survey to the equator. The handling and processing of this mass of data from fragile photographic plates is itself a major operation. By 1985, the UK Schmidt Telescope had taken over 10,000 plates, the telescope time being fairly evenly divided between survey and non-survey work. Master positive glass copies of the Sky Survey plates are made at Siding Spring and sent to the ESO Sky Atlas Laboratory in Geneva, or more recently to the Royal Observatory Edinburgh where film and glass copies of the surveys are made for distribution to over 170 institutions around the world. The plate libraries of the UK Schmidt Telescope Unit at Edinburgh and Siding Spring retain the original plates, and film or glass copies of plates covering all of the southern sky, including those from the ESO Schmidt. Since the UK Schmidt Telescope is operated by a team of resident astronomers, individual researchers requiring special plates can arrange for these to be obtained, and the proportion of time allocated to such work has recently risen to about 70 per cent. The international nature of this activity is shown by the fact that up to March 1985 there had been 201 individual user requests from Britain and 212 from non-British users in 24 different countries.

The first staff resident at the Schmidt included several research astronomers who were among the earliest users of the AAT. From the outset the survey plates provided a wealth of discoveries to be followed up on the larger telescope. In the mid- to late-1970s it was common for regular users of the AAT to arrive for afternoon tea at the Schmidt where they were entertained by a selection of Polaroids depicting the latest celestial treasures. Collaborations blossomed as new planetary nebulae were found and confirmed, and new examples of interesting galaxies explored. Nor were the discoveries restricted to types of objects already in the northern sky: the improvements in emulsions, the matching quality of the telescope, and the dark skies of Siding Spring permitted the detection of much fainter features than previously known. A class of galaxies sparsely populated by stars, and known as low-surface-brightness galaxies came to light. Later, photographic techniques developed by David Malin at the AAO Laboratory revealed even fainter features. Most notable of these were the incomplete arcs of light found around many elliptical galaxies and demonstrated from AAT data to be shells of stars ejected from the galaxies. This discovery has been seminal in the study of elliptical galaxies.

By the 1980s the balance of the Schmidt-AAT projects had swung towards the non-survey plates. The use of the COSMOS and APM plate measuring machines, at the Royal Observatory Edinburgh and Cambridge University respectively, in particular, made possible the identification of rare objects from among the millions of images on direct and prism plates. The search for high-redshift quasars, with its accompanying census of the universe at one-fifth of its present age, when the light left these quasars, has been dominated by the Schmidt-AAT partnership. The number of times that the distance (redshift) record has been broken by the telescopes working together is itself testimony to the co-operation.

At the other end of the distance scale, the Schmidt and AAT also have found some of the nearest stars. The scientific importance of such work lies in the nature of the stars themselves, for they are the very smallest, coolest and least luminous stars known. There previously had been no information on the numbers of such objects, but as the Schmidt-AAT surveys have been extended there is growing evidence that these stars are so numerous that they may account for much of the mass of the Galaxy.

It is clear that the objectives of those who originated the UK Schmidt project have been, indeed are being, fully achieved. The very high standard of research from, and optical performance of, the Schmidt match those of the AAT, and the complementary programmes to which each telescope is suited is by now evident. The case for an all-British out-station at Siding Spring is no longer as strong as it was in the 1970s. It was not surprising that the well established base of the Anglo-Australian Observatory in Sydney and the fruitful co-operation between all the astronomical groups at Siding Spring Observatory prompted establishing an even closer association between the Schmidt and the AAT. In 1988 a formal agreement between the Science and Engineering Reseach Council and the Anglo-Australian Telescope Board marked the conclusion of negotiations for the Anglo-Australian Observatory to assume responsibility for operating the Schmidt Telescope. This agreement has made the AAO a two-telescope observatory headed by one Director under the control of the AAT Board. The events leading to this agreement are described in Chapter 13.

Figure 11.5   Comet Halley taken on 12 March 1986 with the Schmidt Telescope, and printed with an unsharp mask

Notes to Chapter 11

1 Burbidge G. (1972). *Nature* (London) **239**, 119.
2 *Telescopes* (1960). G.P. Kuiper and B. Middlehurst (ed.), University of Chicago Press, Table 1. Of the other Schmidt telescopes, four were in the USA, one in Mexico, and three in the southern hemisphere.

12 Some achievements of the AAT

Once the AAT was completed and handed over in 1974 to its regular operating staff it became part of the astronomical mainstream, and from then on its particular achievements can be put in perspective only by relating them to astronomy as a whole. We will not do that here. Instead, we will give a brief account of its technological development, especially with instrumentation, and describe some of the main scientific achievements of the telescope.

12.1 The technological development of the AAT

When work began on the AAT astronomy was well into the great post-war upsurge referred to in Chapter 1, with notable discoveries coming thick and fast. Astronomers, now brought into much closer contact with physicists, were beginning to realise how remiss they had been in their failure to keep up with current technology, and how much this might affect the efficiency and accuracy of their observations. At the same time, and as a result of the dramatic development in microchips, computers had entered a phase of extraordinarily rapid growth, while various types of electronic radiation detectors, which had been waiting in the wings for years, were surely due to appear on full stage before long. The success of several large radio telescopes, very much bigger than the biggest optical telescope, the 200-inch on Palomar Mountain, had introduced to the optical telescope world new concepts from structural and control engineering. Finally, at least two other groups – those at Kitt Peak National Observatory and the European Southern Observatory – had entered the field, and the frequent meetings and colloquia held with them considerably stimulated ideas on the building of large telescopes. The times were propitious, and the AAT found itself in the forefront of a massive new wave of optical telescope building, a situation of which the Project Office staff was well aware and intended to take full advantage.

The impact of Joe Wampler as first Director combined with the success of the Image Dissector Scanner (IDS) was decisive in a number of ways. It gave the new Observatory an initial momentum it has never lost. Once having used the IDS no observer would willingly obtain data in any other way, and the IDS made remote operation inevitable. But, laid-back Californian that he was, Wampler had trouble with some of the more formal aspects of the Australian life-style. Also, he was a long way from Silicon Valley, and it was no surprise when he returned to the USA in March 1976. Before he left he made two appointments from which the AAO has drawn considerable benefit: those of David Malin, the photographic scientist, and John Barton, the electronics engineer.

Donald Morton, a Canadian who had been at Princeton Observatory for a number

Figure 12.1   Donald Morton and Peter Gillingham in the Cassegrain cage

of years, took over as Director in 1976. He arrived to find a precocious fledgling of
an observatory, staffed with enthusiastic scientists and engineers, but yet to make
its presence felt in wide circles. When he departed, almost ten years later, he left
behind a mature and internationally renowned institution. It is difficult to overplay
Morton's role in this transition. With his eye for detail, dedication to work and acute
awareness of everything that was happening in his observatory, he instilled in his
staff an abhorrence of shoddy and inconsequential work, be it in the maintenance of
the telescope or the finer points of writing reports. Under his aegis the AAO came
to be recognised not merely as the organisation that ran the AAT, but as one of
the foremost astronomical research institutes in the world. It was said of him that
'the uncompromisingly high standard of operation of the AAO which characterised
his term is largely responsible for the fine international reputation enjoyed by the
Observatory'.[1]

Morton also left his mark on Australian astronomy. He served on the advisory
boards of the Sydney Observatory and of the Australia Telescope, the latter then
under construction and now a superbly powerful radio telescope. He was elected
President of the Astronomical Society of Australia; and in recognition of his scientific standing he was made a Fellow of the Australian Academy of Science. These
involvements helped to raise Australian awareness of the AAO.

The early part of his term was marked by an extended struggle for consistently
increased spending on instruments, described in Chapter 13, without which it would
not have been possible for the AAO to have maintained its leading position. One
effective move was to set up the three-man Advisory Committee on Instrumentation

Figure 12.2   From left: Doug Cunliffe, Jean Melville (SERC), Donald Morton and Geoffery Allen, SERC Chairman, at the Epping Laboratory in September 1978

for the AAT (ACIAAT) to examine proposals and make recommendations to the Board. The manpower resources of the AAO being limited, design and manufacturing contracts for new instruments are often let out to university groups, observatories and occasionally to industry. The two Royal Observatories, University College London, the University of Durham, the Australian National University, the University of Melbourne, and the AAO itself have all taken part in this process. In this way the AAO has had access to a much larger pool of innovative instrument designers and builders than could ever be possible in a single observatory.

In Sepember 1986 the Observatory received its third Director, after Donald Morton who had served the Board for almost ten years moved on to become the Director of Canada's Herzberg Institute of Astrophysics. The present Director is R.D. Cannon, an astronomer who came from the Royal Observatory, Edinburgh, where he had been Deputy Director and Head of the Schmidt Telescope Unit. He has continued the policy established by Morton of maintaining a first-class suite of instruments for the AAT.

While pride of place, in the sense that no other instrument is in greater demand, still goes to the RGO Cassegrain spectrograph, some very fine innovative instruments have been built. These include, among others, an infrared grating spectrometer (FIGS), quite the best device in its class when it was made; FORS, the faint object red spectrograph; and Taurus, an interferometric device for measuring the velocity pattern over the whole of a two-dimensional image such as might be produced of an emission nebula or a galaxy. The most recent instrument is UCLES, an echelle spectrograph

Figure 12.3   Russell Cannon, AAO Director, in 1989

built at University College London, used at the coudé focus. Its high resolution, high stability and wide wavelength range extend the spectroscopic capabilities of the AAT quite dramatically. Immediately after commissioning it won about a quarter of all observing time on the AAT. Its essential feature is that it allows astronomers to obtain very high resolution spectra over wide wavelength ranges, conditions which are mutually exclusive in conventional spectrographs. It uses modern, very efficient, panoramic detectors, and has opened up new areas of research for faint quasars and distant stars, as well as re-opening branches of astronomy involving relatively bright stars which have lain largely dormant for many years.

Some of the instrumental developments entirely within the AAO have received world acclaim, and it is appropriate here to highlight two of them: the infrared photometer-spectrometer (IRPS), and the use of optical fibres.

At the outset it was never envisaged that the AAT would be used to any degree for infrared observations. Although a few users were prepared to bring their own equipment to the telescope, they were encouraged to expect a poor site and a telescope that, in the parlance of infrared astronomers, was 'dirty'. At infrared wavelengths objects emit radiation that merely adds to the difficulty of a measurement. All metalwork seen by the detector is emitting unwanted radiation; the mirrors themselves glow dimly; even the sky never gets truly dark. Infrared observing has been likened to trying to see stars in daylight with a luminous telescope. The AAT was designed to baffle out any stray sky light, but the baffles are brighter than the sky at infrared wavelengths. The large secondary mirror supports, and the large hole in the primary mirror all contribute to the telescope's dirtiness.

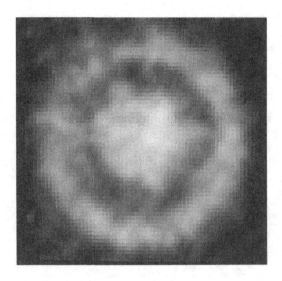

Figure 12.4   An infrared image of the rings of Uranus

However, the infrared is a term used to describe a large portion of the electromagnetic spectrum, and it was recognised early on that the situation was not at all bad at the shorter end of the range, from 1.0 to 2.5 micron wavelength. The IRPS, the first instrument to be developed within the AAO, addressed these wavelengths particularly well. Moreover, a novel design of control electronics by John Barton made the IRPS both unusually versatile and very sensitive. The AAO was the first to offer common-user infrared equipment to Australian or British astronomers, and for ten years it operated the most sensitive system in the world in this wavelength range.

The use of optical fibres was pioneered single-handed by Peter Gray who gave the AAT another first: multiple-object spectroscopy.[2] In doing so he established a technique that has begun to dictate the design not only of instruments but of entire telescopes. The idea is very simple. The light from 50 or more objects, reasonably close to each other in the sky, is ducted along optical fibres to the spectrograph slit so that the spectra of all 50 are registered at the one time. The gain in efficiency is naturally immense, provided that there are sufficient objects of interest accessible at one setting of the telescope. Here the AAT had a clear advantage over its competitors in that the wide field of the f/8 Ritchey–Chrétien focus provides the necessary sky coverage.

The initial experiments with optical fibres have led to other innovations. The light from an extended object such as a galaxy can also be ducted into the linear slit of the spectrograph by fibres. Also, light can be taken from one focus to an instrument at another location, such as on the floor of the dome. This facility both increases efficiency and reduces the number of changes to instruments which are so costly in

Figure 12.5   The multi-object spectrograph. Optical fibres plugged into the locations of galaxies on an aperture plate at the Cassegrain focus of the AAT.

manpower.

Most AAO instruments dispense the light at the output end to a choice of detectors. Since 1977 the basic AAT tool has been the Image Photon Counting System (IPCS), invented by Alec Boksenberg, now Director of the Royal Greenwich Observatory, and built at University College London. The advance the IPCS gave over the IDS was that, whereas the latter could measure data (usually spectral data) only along a line, the IPCS has full two-dimensional capability and can measure every element in a picture that fills the available area. Essentially an enhanced, sophisticated and sensitive television camera, it divides the picture into thousands of elements called pixels, and measures the light intensity at each one several times a second. The sums for the signals at each pixel are accumulated in a computer memory; they can be displayed on a video screen at any time during the observation, and are transferred to a computer memory at its conclusion. Over the years the IPCS has been developed steadily. Its array size has been expanded notably, and it remains significantly larger and intrinsically more accurate than almost any other astronomical detector. The particular strengths are its very low background or 'dark' current, and its good sensitivity at the ultraviolet and blue end of the spectrum.

Recently, however, it has been challenged by the high efficiency and high data rates of the charge-coupled device (CCD). CCDs are simpler than the IPCS, with little of the complex IPCS circuitry. They are made of arrays of tiny light-sensitive microchips upon which the image is focussed directly. A charge proportional to the intensity of the light falling on it builds up on each chip or pixel, to be read off at the end of the observation into a computer memory. CCDs are especially powerful at the red end of the spectrum, and complement the IPCS, although techniques for extending their sensitivity to shorter wavelengths are being developed at the AAO and elsewhere.

All these devices, including those for infrared astronomy, operate near the edge of current technology. Because their use in astronomy where light levels are extremely low usually means that they are being employed well beyond normal commercial specifications, the demands are high on the staff who design, build and operate them. Once again events have proved the wisdom of having a strong electronics section at the AAO where John Barton has earned a reputation for his careful optimisation of CCDs. All CCDs have imperfections, but Barton's electronic circuitry allows the adjustment of the operating parameters so as to minimise those blemishes. Observers at the AAT enjoy some of the highest quality CCDs anywhere.

The role of instrument computers has become steadily more important. As in Wampler's time, they still provide the astronomer with his basic interface to the instrument, control all the instrumental functions, look after data acquisition and handling, and perform at least some of the data analysis. With time the advantages of simplicity and directness in a control program are becoming increasingly clear. The great majority of astronomers who use the AAT are normally on site for only a few days, not enough to become familiar with the programs. When faced with a complex, many faceted machine like, say, Taurus or the coudé echelle spectrograph, it is a great help to find it has a control program which can be mastered quickly and which makes the instrument quite simple to operate. Such simplicity does not come easily. The AAO has four full-time professional programmers, together with six or eight engineers and technicians, all more or less fully involved with computers. Compared to most observatories this is a strikingly large proportion of a workforce of 50; but the result is software of a quality for which the AAO is justly renowned, and which is in many ways the lifeblood of the place.

Nevertheless, the electronic revolution has not totally overwhelmed the AAT, and this is owing entirely to one man, the photographic scientist David Malin. At a time when photographic emulsions were becoming the cinderellas of large telescopes, Malin introduced new techniques to extract information from the plates. His techniques of enhancing faint photographic images have yielded a rich harvest. As but one example, a method of enhancing the visibility of exceedingly faint nebulosity led to a significant discovery that rings of light surround many otherwise normal elliptical galaxies. These have been shown to be produced by concentric shells of stars that have been ejected episodically in some previously unsuspected activity probably associated with the collision of galaxies.

More widely known than his scientific work, however, is Malin's perfection of colour

Figure 12.6   David Malin, photographic scientist, in the prime focus cage

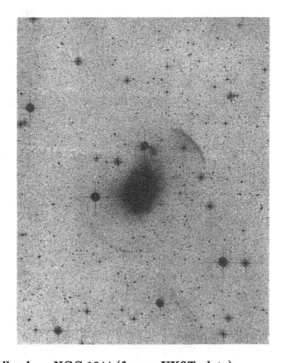

Figure 12.7   A shell galaxy NGC 1344 (from a UKST plate)

photography through the combination of positive copies of plates taken at three separate wavelengths. His colour photographs grace most popular astronomical texts of the last decade, have been the subject of many international exhibitions, and have earned him numerous accolades. They are more than mere works of art. Astronomers can often be seen poring over a Malin print discussing some faint celestial smudge of unusual hue that had long lain unnoticed on monochrome prints.

12.2 Some scientific achievements of the AAT

The scientific staff has been one of the strengths of the AAO. The policy of non-tenured appointments, allowing a constant replenishment of styles and ideas, has undoubtedly contributed greatly to the success, while several have taken full opportunity of the provision allowing up to seven years' employment. The AAO staff are in a privileged position through their intimate knowledge of the telescope and its equipment, and it is no surprise that they have claimed a significant share of the scientific highlights from the telescope.

It is, in fact, a difficult task to identify a manageable number of such highlights, and an invidious one to attach the names of individuals to those few which are selected. Each year the Board's Annual Report identifies half a dozen of them, and the tenth anniversary issue in 1984–85 gave a retrospective view of highlights from the collection. We draw heavily on that selection.

The first example is the Vela pulsar. One of the first to be found, it was discovered in 1968 with the Molonglo radio telescope. Its apparent association with the Vela nebulosity suggested a distance of about 500 pc, near enough to encourage a search for optical pulses. An optical counterpart had been found for the Crab pulsar in 1969, but for no other of the 100-odd pulsars found up to 1977. The task turned out to be difficult. When eventually detected at twenty-fourth magnitude, it was the faintest optical object to have been measured up to its time, and its successful detection gave clear notice that the new telescope had arrived. It was made possible by the close co-operation of the radio astronomers. An accurate position, an essential prerequisite, came from observations made with the Fleurs and Molonglo radio telescopes, and an accurate period, equally necessary, was supplied by Parkes. The initial discovery was made with the two-beam photoelectric photometer, the final confirmatory one with a sophisticated application of the IPCS. As it happened the crucial observations were made on the one night of the year that the Board visited the telescope; on this night it was given the unique opportunity of watching a spectacular discovery unfold at first hand and in real time.

X-ray astronomy came of age early in the 1970s with the launching of the the Uhuru satellite, at about the time that the AAT was approaching completion. Within a few years Uhuru had demonstrated the existence of new classes of objects such as mass-exchanging binaries, active galaxies and quasars, clusters of galaxies, pulsars

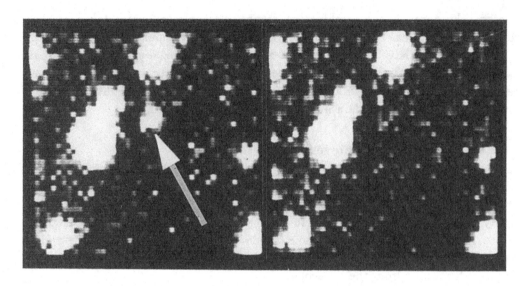

Figure 12.8   Two views of the Vela optical pulsar. Each is the sum of 70,000 television frames obtained during part of the cycle when the pulsar was dim (right), and when bright (left). Each picture is only 12 arc seconds across, and the pulse pattern repeats itself every 89 milliseconds. These images establish precisely which star is the pulse (arrowed), and that it never turns off completely.

and supernova remnants, all very exciting and all crying out for study by optical telescopes. This was especially so for the AAT, because strong X-ray groups had formed at the University of Leicester and University College London, and the British had successfully flown their own X-ray satellite in Ariel-5. Here we can discuss only the star SS 433 (No 433 in the catalogue of Stephenson and Sanduleak), one of the most extraordinary discoveries of its decade. A spectrum of this star taken with the AAT showed peculiar emission features which identified it with a radio source discovered with the Molonglo radio telescope. X-ray astronomers using Ariel-5 then established that the star was also a strong X-ray source, and it was noticed too that it lay at the centre of the supernova remnant W 50. The story was taken up in California, where optical observations showed that the star was a binary with a peculiar spinning disc, along the precessing axis of which two opposing jets of gas are being ejected at speeds a quarter of the velocity of light. Such speeds are totally unprecedented for a stellar system, and the full implications of the result have yet to be worked out.

In Chapter 1 we stated that the existence of quasars alone would have justified the building of a large southern telescope. The AAT's record in this field is in fact quite outstanding, especially when the telescope is considered in conjunction with the UK Schmidt. This is perhaps not surprising when we recall that the pioneer work on radio sources was carried out almost exclusively in Australia and Britain,

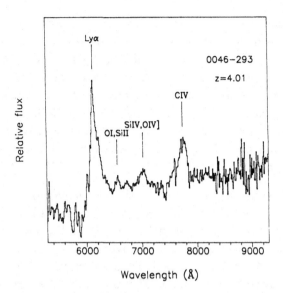

Figure 12.9 The spectrum of the quasar 0046–293, the first discovered at a redshift greater than 4, recorded with the AAT's faint object red spectrograph. The most prominent lines of hydrogen, carbon and silicon are indicated.

and that both countries have maintained a strong presence in the field ever since. Quasars have several claims to importance. Since they are the most distant and the most luminous objects known, they provide the best available tools for exploring the most distant reaches of the Universe or, equivalently, for studying it when it was at a small fraction of its present age. Because they generate such vast amounts of energy within such small volumes, comparable to the solar system in size, they are objects of intense interest. They convey information about otherwise invisible intergalactic objects, such as young galaxies which happen to lie in the line of sight between us and the quasar.

Taken together, the AAT and the Schmidt have been an immensely successful combination for finding large samples of quasars, and several times they have held the record for discovering the most distant examples. Currently distant quasars are most efficiently recognised by their peculiar colours, as measured on automatic plate measuring machines at Cambridge and Edinburgh. Figure 12.9 reproduces the spectrum of the quasar 0046–293 with a redshift which broke the $z = 4.0$ barrier. The Lyman alpha line, the rest wavelength of which is 1215Å, can be seen redshifted to a wavelength of 6092Å. The light left this object when the Universe was less than a tenth of its present age. In another spectacular investigation, 80 candidate objects were selected in one small field; of these, observations with the faint object red spectrograph (FORS) confirmed that 30 were very high redshift quasars, and allowed the first reliable estimate to be made of the space density of distant quasars.

The AAT has been used to study objects at every distance from these remote quasars to the sun's nearest neighbours. Used in its scanning mode with an infrared detector, it has revealed the presence of clouds on the dark side of Venus. At suitably chosen wavelengths the clouds are back-lit by a warm glow from near the planet's surface, and show swirling patterns deep within the Venusian atmosphere. Infrared spectra obtained of comey Halley during the close approach of 1986 showed, quite unexpectedly, evidence for carbon–hydrogen bonds and therefore presumably for the existence of organic material. It is believed the substance involved may be be related to terrestrial tars or coal.

Our final example concerns the abundances of elements in stars. The relative stellar abundances of the heavier elements – oxygen, nitrogen, silicon, calcium, iron and so on – normally mimic those on earth or the sun. From the interstellar medium stars form by gravitational contraction, shine by converting their hydrogen and helium into heavier elements, and die when their fuel is exhausted, to return some or all of their matter to the same interstellar medium. The matter they return, as well as the new stars which form from it, is richer in heavier elements. As a galaxy evolves, we can expect a steady buildup of its metal content together with a broad correlation between stellar age and metal abundance. Abundance determination is therefore important and is studied intensively, not least at Mount Stromlo Observatory. Following indications that its metal abundance was unduly low, two Mount Stromlo observers made a detailed study of the star CD –38°245, to show that its metal content is about 100,000 times less than that of the sun, and at least ten times less than that of the next most metal-deficient star. This figure of 100,000 is unprecedented, and provides some of the best evidence yet that our Galaxy, and by extension the Universe itself, began as a primordial cloud of hydrogen and helium scarcely contaminated by the heavier elements. CD –38°245 therefore occupies a most important place in the stellar hierarchy.

It would be inappropriate to end this chapter without mention of perhaps the greatest scientific highlight that was, as it were, thrust upon the AAO: supernova 1987A. Supernovae, the explosive death certificates of massive stars, occur in galaxies throughout the Universe. At peak light they are as bright as an entire galaxy, but they fade quickly. Supernovae are of interest not only for what they teach us about the evolution of large stars, but because they create all the elements beyond calcium in the periodic table, and most of those between helium and calcium too. Moreover, they are potential 'standard candles' of sufficient luminosity to measure the scale of the Universe to great distances. While common, they are not so common that we expect nearby examples very often. A handful of supernovae each year become bright enough, for a few weeks, to be seen in large amateur telescopes, and quickly fade to oblivion. Most are so faint that few observations of them are ever made.

Because it occurred in the Large Magellanic Cloud, the closest galaxy to our own,

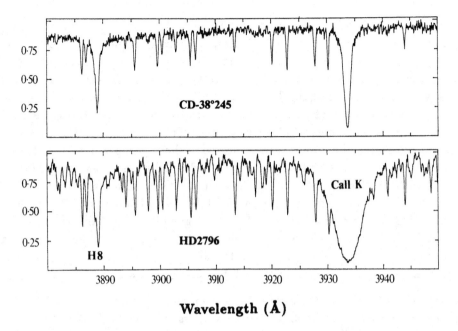

Figure 12.10   Spectra of the ultra metal-poor star CD −38°245, and the next most metal-poor star HD 2796. The strong line at 3933Å is the K line of ionised calcium.

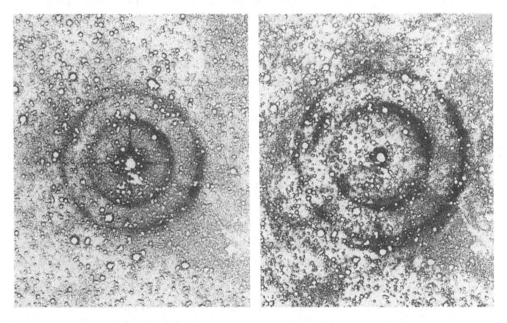

Figure 12.11   The light echoes of supernova 1987A taken with the AAT showing the area of expansion between July 1988 and February 1989

SN 1987A became 10,000 times brighter than most. It was the first supernova visible to the unaided eye for nearly four centuries, and the first since the invention of the telescope. From the moment of its discovery it became the most important astrophysical event in the history of the AAT. Its location in the far southern sky precluded access by telescopes in northerly latitudes, and placed an unprecedented responsibility on the southern observatories to obtain all possibly useful observations, even if the data could not at the time be fully interpreted. The Director, Russell Cannon, immediately recognised the importance of observing the supernova, and quickly secured the agreement of the time allocation panels and AAT users to an override of up to one hour each night when SN 1987A was at its brightest. The AAO has established an archive of these data gathered with the different instruments at its disposal. A new, ultra-high-dispersion spectrograph was designed by Peter Gillingham, and built essentially from spare parts, some of them wooden, to make unique observations of the supernova. This instrument illustrated an in-built Australian ability to make things work, and was an example of bush technology in its highest form.

As the supernova faded, the data became even more important. SN 1987A was the first for which data could still be taken several years after the initial outburst, allowing astronomers to study, as never previously done, the late evolution of a supernova. The collected data, to which the international community has access, will remain a benchmark at least until the next such event, and potentially long beyond then since supernovae are by no means identical. In contrast to the normal fickle progress of science, where data from new techniques supersede old, these data will endure for centuries.

An unexpected benefit of archival data, taken of the region before the supernova exploded, has become evident with the latest manifestation of the exploding star. Minute particles of dust near the line of sight to the supernova scatter light towards us. Because the path length is slightly longer than the direct route, we see echoes of the light which left the supernova near its maximum, but delayed by the extra distance travelled. By far the best images of this very faint effect have been obtained photographically, by subtracting images on recent AAT plates from similar plates of the field taken several years ago. In this way the complicated nebulosity in the region is cancelled and the light echoes revealed as complete circles literally reflecting the structure of the elusive and normally unseen interstellar medium. As the echoes continue to expand the archival plates will be used repeatedly to map and to study the interstellar medium in the region of the supernova.

It is in such unexpected ways and from such surprise events that scientific advances are made. Nobody could have predicted that SN 1987A would have appeared, or that archival plates of the region would have proved so useful. So it is with the suite of observations collected since February 1987.

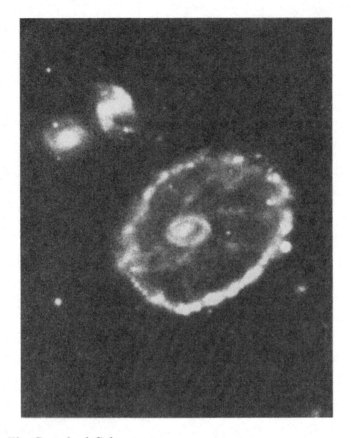

Figure 12.12   The Cartwheel Galaxy

Notes to Chapter 12

1 AAT Annual Report 1985–86 p.1
2 Hill J.M. (1988). The History of Multi-object Fibre Spectroscopy, *Astron. Soc. Pac. Conf. Series* **3**, 77.
  See also Warren S.J. *et al.* (1987). *Nature* (London) **330**, 453

13 Political winds of change

After the AAT's detailed commissioning period, regular scientific work and scheduled observations began on 28 June 1975. By 1977, the Board had substantially completed the major construction phase for the telescope, and had established a permanent laboratory in Sydney complementary to its facilities at Siding Spring. The Board finally closed its construction account at the end of the 1976–77 financial year, and recorded the total amount of $15,932,250 paid by the two Governments towards the project. The telescope essentially was entering its next phase.

The Board membership, too, was changing along with the staff at the Observatory. In 1977 the three British members were Sir Harrie Massey, Professor V.C. Reddish and Mr M.O. Robins. Massey's association with the AAT went back at least as early as his meeting with Senator Gorton in 1966 when the idea for a large southern hemisphere telescope was still maturing into a bi-national project. Vincent Reddish was Astronomer Royal for Scotland, and as Director of the Royal Observatory, Edinburgh, had close links with Siding Spring through the Schmidt Telescope. Mac Robins was the Science Research Council's representative, and in turn he would be replaced by Dr H.H. Atkinson. The Observatory was truly an observatory now, and in the previous year it received its second director, D.C. Morton. As the work of the Observatory expanded, the Board separated the tasks of Secretary from those of Executive Officer, both of which Doug Cunliffe, the Executive Officer, had been performing. In 1974 the Board had acquired its own Secretary on secondment from the Australian Government agency under a fixed-term appointment to attend to policy matters which were the province of the Board itself. The first Secretary under these arrangements had been George Kazs, followed by Kevin Bryant three years later.

In the following financial years the Board's tasks now were operating the facilities in Sydney and at Siding Spring, and maintaining and developing the AAT's suite of instruments at the highest prevailing international standard created by the revolution in instrumentation in the mid-1970s.

It was the British Board member Jim Hosie, in the light of his experience with the Isaac Newton Telescope and British space programmes, who ensured that there should be on-going funds for development of AAT instrumentation. The wisdom of this initiative was apparent by the time the telescope was completed. The Joint Policy Committee in March 1969 had before it a proposed budget for instrumentation of $500,000, and it began to set priorities for basic common user instruments which included cameras, Cassegrain spectrographs, photometers and a high resolution coudé spectrograph. However, it was soon evident that the proposed budget

Table 13.1. *Funding by the two Governments since 1974–5, with the percentage increase or decrease on the budget of the previous year*

1974–5	1975–6	1976–7	1977–8	1978–9	1979–80	1980–1	1981–2
			$A '000				
1951	1632	1660	1700	2080	2200	3060	3476
–	–16.3%	+1.7%	+2.4%	+22.3%	+5.8%	+39%	+13.6%

1982–3	1983–4	1984–5	1985–6	1986–7	1987–8	1988–9	1989–90
			$A '000				
3768	3670	3590	3794	4002	4766	5046	5348
+8.4%	–2.6%	–2.2%	+5.7%	+5.5%	+19.1%	+5.9%	+6%

Table 13.2. *The annual instrumentation budget and the percentage it formed of the whole annual budget*

1974–5	1975–6	1976–7	1977–8	1978–9	1979–80	1980–1	1981–2
			$A '000				
–	66	80	92	212	110	266	364
–	4%	4.8%	5.4%	10.2%	5%	8.7%	10.5%

1982–3	1983–4	1984–5	1985–6	1986–7	1987–8	1988–9	1989–90
			$A '000				
358	240	545	692	431	680	813	925
9.5%	6.5%	15.2%	18.2%	10.8%	14.3%	16.1%	16.5%

fell considerably short of what the Board wanted for the telescope, and the budget was increased to over \$1 million later that year. In 1974, to extend the scope of the original instrumentation the Board embarked on a major programme of instrumentation development over four years. The total cost of this development between 1974–5 and 1977–8 was estimated at \$1.5 million. Added to this was the need for both a scientific headquarters and adequate accommodation at Siding Spring, which increased the four-year estimate by a further \$925,000.

13.1 Winds of change in Australia

The Australian Government convened a committee, known as the Westfold Committee[1], in 1976 to recommend policies for the co-ordinated development of the various areas of astronomical research and for the rationalised use of astronomical facilities in Australia. The Government also asked it to examine ways in which savings might be made in the Government's funding of astronomy. The Committee's report, *Review of Observatories*, was published in 1978. It showed that in 1976–77 approximately \$8.7 million had been spent on astronomical research in Australia, and of this \$7.6 million was provided by the Commonwealth Government. The remaining \$1.1 million came from the British Government's contribution to the AAT, and from the State Governments. While this report did not affect directly the AAT, one of its results was the establishment of an Astronomy Advisory Committee in July 1979 to advise the Minister on matters concerning astronomy in Australia. This Committee conducted an important review of the AAT's future resources in that year.

In early 1978 the Board held a workshop on instrumentation as a necessary step towards the evolution of a long-term plan for providing instrumentation for the AAT. Such a plan was necessary to determine the order of priority for building and procuring a range of instruments under conditions where financial and human resources were limited. A working group[2] appointed by the Department of Science produced recommendations for new instruments which would allow for the most efficient use of the AAT and its existing instruments. The document[3] was seen as a powerful source of advice to the Board, and one on which the Board could base a strong argument for future instrumentation funding. As a result of this review, the Board established its own permanent Advisory Committee on Instrumentation for the AAT (ACIAAT) in September 1979 which reports to it annually.

In the late 1970s the Australian Government faced an acutely difficult budgetary situation. During this period it established a new Department of Finance to stand alongside the Treasury and provide a second voice on matters concerning the economy. There was a policy of 'no real growth' in the public sector, and Cabinet had issued specific expenditure guidelines to all Departments. As part of this policy Prime Minister Fraser set up the so-called Lynch Razor Gang to identify and to implement cost cutting measures in all areas of public spending.

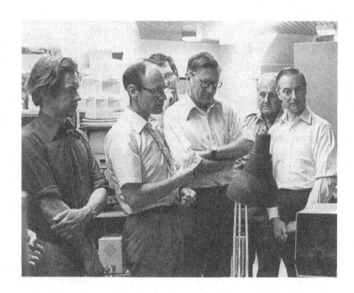

Figure 13.1   Board members from left: Harry Atkinson, Donald Morton (AAO Director), Paul Wild, John Farrands and Vincent Reddish in the AAT control room in March 1979. Partly visible behind them is John Carver.

It was against this background that the Board approached both Governments for a total budget of \$2.4 million for 1979–80. The Minister for Finance had put forward proposals to limit demands on public funds, and considered that support for the Board's case for funding could be seen as preferential treatment.[4] Following extensive consultations, including with the Department of Science and the Environment, the Australian Government indicated that it was prepared to support a half-share contribution which, if agreed by the United Kingdom, would have the effect of limiting the budget to \$2.02 million. Such an allocation would represent a small real increase in funding, and would allow the Board to initiate work on the dome seeing problem and to obtain additional computer equipment.

When Sir Harrie Massey, Deputy Chairman of the Board, was visiting Australia in February 1979, he and the Board's Chairman, Paul Wild, held a special meeting with representatives of the Australian Government. Here Sir Harrie presented the British case for maintaining the higher level of funding, and explained that the British community regarded the AAT as a very successful project, and made the point very strongly that this was largely because of the AAT's instrumentation and associated facilities. He emphasised that the Board's budget was necessary not only to maintain the telescope but to ensure its instrumentation and computers kept pace with the latest technology. To achieve the latter, it was vital to start early rather than to defer projects. Massey and Wild stressed that if the AAT were funded at the low level offered for 1979–80 the telescope would fall behind and possibly lose its world-

Figure 13.2   Board member (1975–83) Sir Harrie Massey, FRS *(Photo: UCL)*

class position in astronomy. In fact, in 1978–79 the AAT's budget had been lower than the SRC wanted to see. For the next financial year, despite a reduction in the funds available for science in Britain in real terms, the SRC was prepared to increase its contribution to the level sought by the Board. The SRC expressed its strong concern and indicated it was willing to take up the matter at ministerial level.

Sir Harrie mentioned specific matters of concern to British astronomers. First, since the average seeing conditions in Australia were not as good as at other major sites, it was necessary to ensure that any additional problems arising from poor seeing in the dome were removed. Second, astronomers travelling from Britain were frequently on a very tight schedule with teaching commitments at their home universities, and preferred not to have to visit Epping. Therefore, they wanted facilities at Siding Spring to be as good as possible. In particular, the British wanted the computer facilities there to be given a very high priority. Sir Harrie concluded that, taken in conjunction with previous cuts to the Board's budget which the British side had reluctantly accepted, there would be many problems if the budget were reduced to $2.02 million. Indeed, the SRC regarded $2.2 million as very close to rock-bottom, and had increased its support for the AAT over a period in which its own total budget had been falling.

The Australian Government's view remained unchanged: there was no argument why the AAT should receive special treatment in the difficult final years of the 1970s. However, the Department of Finance proposed that because it was a long time since the Australian Cabinet had considered the AAT project, the Board should prepare a paper on the future funding of the telescope. If such a paper were placed before the Cabinet, it would allow the Minister for Science and the Environment and the Minister for Finance to reach an understanding on an indicative level of funding for the Board over a period of some years beyond 1979–80. Accordingly, a five-year programme for the AAT was drawn up with the intention of incorporating it in a Cabinet Submission to the Australian Government.

In the meantime, the Board renewed its request for an Australian contribution of $1.2 million in 1979–80, but in April the Minister for Finance agreed to no more than $1.1 million. The Chairman, Paul Wild, had no alternative but to advise the Board to accept this, although it left the AAT with a budget lower in real terms than its previous year's funding. Not surprisingly, the Board and the SRC were unhappy with this result. The SRC representative on the Board, Harry Atkinson, reminded his opposite number, John Farrands, that this was the fourth year in which the Australian side had imposed budgets lower than those the Board recommended and which the British side had endorsed. He said that if increased budgets for subsequent years were not available the whole enterprise would be in jeopardy, and that a long-term solution must be found to ensure the project's stability at a satisfactory level for the future. Atkinson proposed a reaffirmation by the two Governments of their willingness to support the AAT as a world-class facility. He expressed concern at any suggestion of adopting unequal funding because such a move would obviously require consideration of a corresponding change in the share of observing time. The British view was that the AAT should remain fully a joint venture, and a departure from this would undermine the essence of the collaboration between the two national communities.

By August 1979 an expert review committee was set up comprising D.C. Morton, the AAO Director, Professor M. Disney (nominated by the British SRC) and Professor D.S. Mathewson (nominated by the Australian DSE). The committee's terms of reference were to identify the minimum level of resources to maintain the AAO as a world-class astronomical observatory over the next five to ten years, with special reference to the requirements for the development of instrumentation and facilities; to examine further the consequences and benefits of providing resources above this level; and to report to the AAT Board. The Board regarded the Report of its Instrumentation Workshop of March 1978 as the obvious starting point for this new review. In October 1979 Farrands asked the Australian Government's Astronomy Advisory Committee to advise the Government in the review of future resources for operating the AAT and developing its instrumentation. The Committee strongly supported the view that increased funding was necessary to enable the AAT to improve its instrumentation, and thereby maintain its reputation as a centre of scientific excellence.

The Board, principally through its Chairman, Paul Wild, and the Secretary, Kevin Bryant, prepared a submission to both Governments based on the findings of the

Table 13.3.   *The recommendations in the Cabinet submission for funding for the*
               Board's five-year forward look period to 1984–85

	1980–1	1981–2	1982–3	1983–4	1984–5
			$A '000		
Salaries and general administration	1915	1935	1960	1950	1950
Instrumentation – minimum	630	810	610	490	590
Instrumentation – optimum	910	1080	610	610	540
Building	310	310	485	415	315
Total minimum from each Government	1415	1515	1515	1415	1415
Total optimum from each Government	1555	1650	1515	1475	1390

expert review committee and the Astronomy Advisory Committee. This formed the
basis of a Cabinet Submission drafted by DSE which described the role of the AAT in
Australian and British science, the scientific objectives of a ground-based optical telescope, the AAT's scientific achievements, a comparison between the AAT and other
observatories, its operating requirements, and it funding priorities. The submission
demonstrated that astronomy is a competitive science, and made recommendations
for funding for a five-year forward look period, as shown in Table 13.3. At this time
the Minister for Science and the Environment was Senator J.J. Webster, a Minister with a high profile who was identified with the Liberal business community of
Melbourne. However, he left politics in December 1979 to become High Commissioner for New Zealand, and his replacement was David Thomson, a politician from
the Liberal–National Party Coalition in north Queensland, who was not perceived as
having the same influence in the Cabinet room as Webster. Some members of DSE
were concerned that this change in Ministers could effect the smooth passage of the
Board's vital submission. One of Thomson's first tasks as Minister was to present
to Cabinet *The Future Development of the Anglo-Australian Telescope* at the end of
1979. Indeed, he did so very ably, and the Cabinet accepted it to the Board's relief.

The expert review committee regarded the AAT as already losing its position in
world astronomy because the Board's budgets had been reduced in real terms between
1975–6 and 1979–80. Consequently, it recommended an accelerated instrumentation
programme supported by the optimum line of funding shown in Table 13.3. In particular, this funding would allow for development in the area of solid state detectors,
which were emerging as a new and important field, and would take the AAT into the
next generation of instrumentation.

As a result of the successful passage of the submission to the Cabinet, the Board could proceed to argue for more funds than would accrue on the normal basis that the Government was prepared to distribute to departments and authorities. In effect, there was a case for the AAT receiving special treatment at a time when the fight for funds was very tough. Shortly afterwards the Australian and British Governments agreed on a level of funding increased by 25 per cent for the AAT over the next few years.

Paul Wild's contribution to winning the new level of government funding for the AAT had been vital. On Wild's retirement from the Board in 1982, Sir Harrie Massey described him as having that special ability to temper the cold light of reason with a warm outlook for people and for science. In addition, he had played an important part in negotiating the arrangements with the Australian National University which produced the present fruitful co-operation with the Board.

Until this time the Board's Secretary continued to be an officer seconded from the Australian Government agency, the Department of Science and Technology. This arrangement had been part of the administrative support supplied by the Government in the Board's early days. In 1982 Greg Tegart, the new Secretary of the Department, explained that this arrangement effectively was depriving his Department of a valuable staff position which it could no longer afford in times of governmental cost saving measures. He proposed ending the arrangement when Max Franklin finished his term in 1983. Thereafter, the Board employed directly its own Secretary, and Katrina Proust became the first to take up the new position at the AAO Epping laboratory in that year.

13.2 Winds of change in Britain
If the end of the 1970s produced a difficult economic climate for astronomy in Australia, it was the 1980s, and especially the middle years from 1983 to 1987, which appeared as a tough period for ground-based astronomy in Britain. After fighting and winning the battle for funding from the Australian Government, only a few years later the Board was faced with pressure from the British side to examine its budget closely, and to make cuts where these would not affect the quality of the science. Throughout the difficult budgetry period in Australia the SRC had maintained pressure on the Australians for a healthy budget for the AAT, and the British Board members stressed that the AAT had to be funded as a front-line international facility. However, in the early 1980s the effects of the British Government's fiscal policy filtered through British science policy to astronomy and to the AAT itself.

In April 1981 the SRC changed its name to the Science and Engineering Research Council (SERC), recognising a shift of emphasis towards the support of basic and strategic research which had been taking place for a number of years. In the early 1980s inflation began to exceed the allowances made for it within the British Govern-

Figure 13.3   The Board at University College London in 1985. Back row from left: John Carver, Robert Wilson, Greg Tegart, Alfred Game (SERC); front row from left: Katrina Proust (Secretary), Bob Frater, Harry Atkinson, Malcolm Longair and Donald Morton (AAO Director) *(Photo: UCL)*

ment's allocations to the Council. The SERC's balance of resources would be affected seriously by the complex problem of exchange rate variations on the cost of its international commitments. Although these effects were great and outside its control, the Council had to accommodate them within its existing budgets. By the early 1980s British science was in the position of having insufficient funds for a properly balanced programme, and its funds were thought to be well below the level justified by the scientific opportunities.

The first obvious signs of pressure on funds appeared in 1981–82; the Council could not allocate resources for capital projects to proceed as quickly as it wished, and the competition for research grants was fierce. When the AAT Board members met in September 1981 and March 1982, Harry Atkinson said that as Britain completed its ground-based astronomy projects in the later part of the 1980s there would be a shift of emphasis to space-based astronomical research. Britain wanted to take a lead in new and exciting space projects in which it had previously only been a supporting partner.

In 1983–84 the deficit on international subscriptions amounted to £4.3 million. This was caused by a number of technical factors including higher levels of inflation abroad, a lower value of sterling and an increase in the British share of those budgets where contributions were based on a relative calculation of the GNPs of member states. The British subscriptions to CERN and the European Space Agency made the greatest demands on SERC resources. Within the British Government's cash limits system the Council had to make cuts to its domestic programme. Accordingly, it therefore reduced the working allocation of each of its four boards by £1 million, and produced forward look figures to 1986–87 which indicated that the decline in real terms in its budget was likely to continue. These measures left the investment in astronomy at a low level, and the opinion in Britain was that the AAT should not be exempt from cuts which other programmes were forced to accommodate. In March 1983 it was uncertain whether the British contribution to the AAT's budget for 1983–84 would meet the amount sought by the Board. In the end, the Board froze expenditure for certain items until later in the year when the SERC could assess the competing claims for funds after fluctuations in exchange rates, and their effect on its international commitments were known. By the next Board meeting six months later, the adverse fluctuations in exchange rates and insufficient compensation for overseas inflation meant that the SERC might have to find as much as £8 million per year in savings from programmes in the Astronomy, Space and Radio (ASR) Board. Since the British funds for the AAT came from the ASR Board's allocation, Atkinson announced that the SERC was prepared to unblock only half of the 10 per cent increase for indexation in all the AAT's budgets in the forward look period after 1984–85.

Although it would be necessary to trim the AAT's budget, the SERC in no way wished to reduce the scientific value of the telescope. Nor did these measures contradict the widely held British view that the AAT was very good value for the British community's investment in it. In 1984 the ASR Board reaffirmed its intention to complete its current investment in ground-based projects and to place increased emphasis on space astronomy after 1985. To this end the British side would be looking to reduce the AAT's budget by £50,000 both in 1985–86 and 1986–87. However, it did not intend running down the telescope and reducing its effectiveness, and would seek to make these savings through improved efficiency or new operating methods.

As the demand for research funding continued, it was in 1985 when these pressures would come closest to threatening the AAT itself. British astronomers were preparing for a major event in European science with the inauguration of the Observatory on La Palma in the Canary Islands in July. This observatory in the northern hemisphere had been mooted about the same time as the AAT, and it was ironic that a highlight of British astronomy and science, such as this in 1985, should have taken place against a background of constraints to scientific spending. It became clear that within its ground-based astronomy programme, the SERC gave highest priority to the maintenance and development of its northern hemisphere facilities on La Palma and Hawaii.

In the SERC's five-year forward look plans it had identified a core of highest priority programmes. However, since the ASR Board had insufficient funds to include all the current ground-based astronomy programmes, it decided to provide sufficient funds for the facilities in the northern hemisphere to be properly developed and exploited. This meant that the overall provision for the southern hemisphere would have to be reduced by more than half by the year 1990. Southern hemisphere astronomy comprised the AAT, the UK Schmidt Telescope and the South African Astronomical Observatory. As a first move toward these savings, the SERC decided to give one year's notice of its intention to suspend the agreement with South Africa from mid-1986. But this presented savings of only £350,000 per year, and more savings had to be made. Speculation grew over the fate of the AAT. There was fear that Britain might try to abrogate its commitments under the AAT Agreement, or to renegotiate it. Support both for the AAT and for preserving British access to the southern hemisphere was strong. Yet there was debate even among astronomers whether they needed observatories in both hemispheres in order to do first-class astronomy.

Quickly, members of the astronomical community in Britain drew up a petition calling for protection of the AAT's funding. Within 24 hours the names of 175 astronomers from all over the country had been gathered using the Starlink computer network, and the petition was presented to the SERC. To be effective, the petition had to speak for those who would be using the AAT in the 1990s. Robert Wilson, the Deputy Chairman of the AAT Board, saw the wisdom in presenting in overwhelming numbers the voice of astronomers under 40 years of age. Indirect support, too, for the AAT came from the Space Telescope Advisory Committee which held that there was a vital need for a 4-metre class telescope in each hemisphere.

In his 1984–5 report to Parliament, Sir John Kingman, Chairman of the Science and Engineering Research Council, wrote:

> It is however one thing to build a major facility, and another to instrument and operate it effectively. It is difficult to see how full value is to be gained from the investment of public funds in these facilities unless more resources can be found for their exploitation.
>
> The truth of the matter is that the UK is trying to maintain its science base on the cheap. If we are as a nation to maintain the prosperity and the standard of living which we expect, we need to make the most of our scientific and technological talent. Investment in basic science is investment in the long-term future of the nation, and an inadequate level of that investment will ensure our eventual decline.[5]

On the AAT Board itself, the Australian side was unwilling to reduce the current level of funding. Nevertheless, under the terms of the AAT Agreement one side could not force the other to match agreed contributions. While one side might wish to pick especially as a reduction in contributions was mirrored in a reduction in observing time for that party. Changes as drastic as these amounted to renegotiating the AAT Agreement. Informally a number of options were considered, such as Australian involvement in the SERC's spallation neutron source. But at its meeting late in 1985 the Board itself gave particular consideration to the relation of the AAT to the Schmidt Telescope on Siding Spring Mountain.

The Board was unanimous that the telescope must remain competitive, and it recognised that the current instrumentation programme provided a suite of instruments to hold the AAT in its world-class position. That the AAT had escaped major curtailments to its instrumentation programme in the past was one of its strengths now. There was no dispute that the AAT represented very good value for money in British and Australian astronomy, and much of the credit for this had to go to its Director, Donald Morton.

13.3 The resolution

Science budgets in Britain were again seriously constrained in 1985–86, and this situation was made worse by the wayward behaviour of the international currency markets and the poor performance there by sterling. The display of support for the AAT in August 1985 by British astronomers was very important in showing how highly the telescope was regarded but it did not solve the financial problems. An alternative way of making savings in the southern hemisphere for the SERC would have been to close the Schmidt Telescope. Although an entirely British operation at Siding Spring Observatory by the Royal Observatory, Edinburgh, it was seen as a significant part of Australian astronomy, and its loss would be felt by both parties. In 1986 the SERC asked the Australian side of the Board to examine ways of achieving a closer association between the AAT and the Schmidt Telescope, which would provide savings in the SERC's southern hemisphere astronomy budget. A merger of the two telescopes was on the agenda.

During the 1980s the Australian Government had provided large funding for radio astronomy in the Australia Telescope project which was nearing completion. In addition, the Australian National University had its own advanced 2.3-metre instrument at Siding Spring Observatory, and the astronomers at the University of Sydney were soon to receive healthy funding for a new interferometer at Narrabri. Astronomy had fared very well. As negotiations about a merger of the AAT and the Schmidt Telescope developed, the Australian side of the Board was cautious of any arrangement which might require considerably more money for astronomy. Indeed, Government policy was not to increase its present financial level of involvement in Australian astronomy. But the Board was well aware that measures had to be taken, acceptable to the Australian Government, which would protect the AAT.

The model for which the Board agreed to pursue funds was one where the Anglo-

The model for which the Board agreed to pursue funds was one where the Anglo-Australian Observatory would subsume the Schmidt Telescope to operate and to fund it under the same principles applicable to the AAT. The AAO in effect would become a two-telescope observatory under a single director answerable to the Board. An important factor in this was the arrangement for the use of support facilities at the Royal Observatory, Edinburgh, which formed an integral part of the Schmidt operation, and were essential for exploiting the science from the telescope. In July 1986, the Australian Department of Science was discussing directly with the SERC the arrangement proposed by the Board. The SERC's ASR Board had endorsed the proposals and was keen to see negotiations proceed. Unfortunately, financial constraints prevented Australia from moving towards an agreement on a share of British facilities for neutron beam research, and negotiations were principally for arrangements in astronomy which an amendment to the AAT Agreement could accommodate.

One of the British Board members since 1983 had been Malcolm Longair, Astronomer Royal for Scotland and Director of the Royal Observatory, Edinburgh, which operated the Schmidt Telescope Unit for the SERC. When many astronomers in Britain were against proposals to reduce the position of the Schmidt in Britain's astronomy programme, he recognised the need to give up some component of that programme in order to protect others in it. The Board was convinced that the future of southern hemisphere astronomy for both communities could not be assured unless it accepted a new model for the AAO as a fully integrated two-telescope observatory. In the light of this understanding, Longair fought for an adequate budget to enable the AAO to operate both the AAT and the Schmidt effectively, and to maintain the scientific excellence of each telescope. His support for the new arrangement allowed the substantive matters of the agreement to proceed smoothly.

By this time, the lawyers on both sides were discussing the legal ramifications of the Board's decision. The AAT Agreement limits the Board's powers and functions solely to the operation of the AAT, and to enable it to operate both the AAT and the Schmidt would require an amendment to the Agreement. However, the Australian Government lawyers advising the Department of Science proposed that an amendment to the Agreement should be avoided because the procedure involved was slow and complex, requiring Australian Cabinet approval. There was a simpler and equally effective solution which operated under Australian law. The legal instrument which gives the Board existence under Australian domestic law is the Anglo-Australian Telescope Agreement Act, 1970, and the Australian Parliament can amend the Act to provide the Board with additional powers and functions as may be conferred by Regulations under the Act. Of course, the British Government would be consulted about the details of such a change to the Board's position. It was proposed that such additional powers and functions would include the power to operate another telescope. Although the procedure for promulgating a regulation involved the Governor-General

in Council, it seemed the best option for a speedy solution. The Australian lawyers
set to work drafting the necessary amendments to the Act.

The pressure from the British side to proceed as quickly as possible increased as negotiation stretched from one financial year into the next. The SERC was eager to set
the new arrangements in place, and was organising its financial programmes accordingly after 1987–88. The Australian Minister for Science, Barry Jones, wrote to the
British Secretary of State for Education and Science, Kenneth Baker, in April 1987,
supporting an amalgamation of the AAT and the Schmidt Telescope. He acknowledged the financial stringencies on the SERC, and that its priority was to strengthen
northern hemisphere astronomy at the expense of its programmes in the southern
hemisphere. However, he was eager to see the UK–Australian co-operation in optical
astronomy continue as a significant component of Australia's basic research activity.
Jones outlined the proposal to effect the amalgamation by the simpler procedure of
amending the AAT Agreement Act, rather than the formidable task of an amendment
to the Agreement itself.

In the meantime, the Australian side still had to argue the case for additional
funding, and this became more difficult as time passed. The Department of Finance
interpreted the proposed merger arrangements as new policy, as distinct from continuing policy, and maintained that Cabinet approval was required. To expedite matters,
Jones wrote to the Prime Minister in May 1987 seeking his approval of the merger
outside the normal governmental process for new policy. The Prime Minister, Bob
Hawke, agreed to an additional $250,000 being applied to the administrative costs
of the Schmidt Telescope provided that the money was found as an offsetting saving
within the Science portfolio for the 1987–88 financial year. However, the prospect
of a federal election in Australia was increasing, and Tegart impressed on the Board
in March 1987 the need for speed to avoid additional delays and complications if an
election occurred in the middle of the year.

The Australian Constitution states that the Federal Government will call an election not less than every three years. This is a short term by the standards of many
western countries, and a source of much debate in Australian constitutional circles.
In effect, it has led to very few Governments having completed their full terms.
As expected the Australian people were again called to elect a new Government in
June 1987. When the Hawke Labor Government was returned, it immediately set
about a major restructuring of the Commonwealth administration. In particular, it
reduced the previously existing 27 departments to 17 super-departments, abolished
the Science portfolio, and divided its functions among four departments. The new
Department of Employment, Education and Training (DEET) became responsible
for the Anglo-Australian Telescope Board. With the demise of the Department of
Science, Tegart became Secretary of the Australian Science and Technology Council, a body created in the late 1970s to advise the Prime Minister on science and

Government agency, did not bode well for progress with the arrangements for the merger of the Schmidt Telescope with the AAO. The Australian scientific community was shocked and alarmed because it saw such major changes in the country's administration of science as degrading its importance.

Before the election, the Australian lawyers had made good progress with legal matters, but the legislation necessary to amend the AAT Agreement Act had been caught half-way between the two Houses of Parliament, and the whole legislative procedure had to start again when the new Government was in place. The legislation was finally passed through the House of Representatives and the Senate in September. Of more concern was the interruption to the progress on funding the merger. Before the election Tegart had identified off-setting savings from other areas in his Department to support the merger. Now these sources of savings disappeared along with his Department. Tegart was able to convince Dr V. Fitzgerald, the Secretary of DEET and a keen amateur astronomer, of the strength of the case for the merger, and in the budget negotiations Fitzgerald's strong briefing of his Minister, John Dawkins, enabled the Minister to win the additional $250,000 on a continuing basis for the merged Observatory. But there was an important proviso: the first year's funds had to be used before 30 June 1988. The Board agreed the merger should take effect from 1 January 1988. It was now a matter of settling the terms of the Regulation, and of finalising the Memorandum of Understanding which the Board and the SERC would sign as the instrument to give legal effect to the arrangements. The Board hoped to have these matters in hand by the next meeting in Australia in March 1988.

A further complication in the merger was the issue of the lease between the Australian National University and the SERC for the site occupied by the Schmidt Telescope at Siding Spring. This needed amendment as a result of the merger. With the co-operation of the University and the strong support of the Pro-Vice Chancellor, Ian Ross, an agreement was reached whereby the SERC nominally retained the lease and passed custody of the telescope to the Board.

All believed the merger was on track again when a further obstacle arose. The British legal advisors had previously queried proceeding only on the basis of an amendment to the Australian Act of Parliament. They now asserted that an amendment to the AAT Agreement itself also would be necessary. While amending the Australian Act would change the powers of the Board under Australian domestic law, they maintained that the Agreement needed to be amended in order to enable the Board to use those powers. Renegotiating the Agreement, a document concluded at treaty level, would require the approval of the Australian and British Governments, and thereby introduce further long delays. However, the British lawyers subsequently proposed that the amendment could be achieved by a treaty level Exchange of Notes between the two Governments. The Australians were concerned that even this could take several months so that, with the additional time required for the legal formalities of approval by the Governor-General in Council, new funds provided by DEET could not be used in the current Australian financial year. This would make it extremely difficult to argue for new money, and the merger then would not proceed.

The Australians proposed that, since the SERC was retaining ownership of the Schmidt, and placing it in the custody of the Board, there was no need for a treaty-style document; a simple, non-treaty exchange of minutes would suffice. The Board and the SERC then could sign a Memorandum of Agreement, and the Australian Government could proclaim appropriate Regulations to enable the Board to operate the Schmidt under the Act. This procedure could be carried out within the short time frame now remaining. However, the British lawyers would not accept this proposal, and a deadlock developed.

When the Board next convened in Canberra in March, Tegart had arranged a meeting between Board members and Minister Dawkins. It had seemed an appropriate way to mark the strengthening of Anglo-Australian scientific collaboration, and preparations were in place well in advance for dinner with the Board, the Minister and senior officers of his Department in the evening of 24 March. In addition, the Board's visit to Canberra was planned to coincide with a special scientific meeting of the Royal Astronomical Society at the Academy of Science for the Australian Bicentenary.

As so often happens, dinner tables rather than conference tables are where the most difficult problems often are resolved. Minister Dawkins strongly stated his desire for the amalgamation to proceed speedily, and thus enable the $250,000 for which he had fought in Cabinet to be used before 30 June. Atkinson agreed to convey these views to the British side, and after further discussions on his return to Britain, the British lawyers agreed that the merger could be effected by an Exchange of Letters between the two Ministers concerned. In early April, Dawkins wrote to his British counterpart, Baker, in these terms and the deadlock was broken.

The Board agreed that the amalgamation should take effect retroactively from 1 January. It was now a matter of finalising details, which the lawyers concluded over the following weeks against a tight deadline. At the Royal Observatory, Edinburgh, on 15 June 1988, Bill Mitchell, Chairman of the SERC, and Robert Wilson, Chairman of the AAT Board, signed the memorandum of agreement whereby the Board became responsible for the care, maintenance, operation, management and development of the Schmidt Telescope. The agreement also described the access which scientists would have to the fast measuring machines and associated facilities in Britain.

The Anglo-Australian Observatory, now more than 14 years old, had become a two-telescope observatory by a margin of 15 days.

Notes to Chapter 13

1 Membership was K.C. Westfold (Chairman), O.J. Eggen, S.C.B. Gascoigne, R. Hanbury Brown, M.D. Waterworth, D.C. Morton, J.P. Wild, V.D. Hopper and I.R. Ross.

2 Membership was A.R. Hyland (Chairman), K.C. Freeman, D.L. Jauncey, A.W. Rodgers, R.R. Shobbrook, J.A. Thomas.

3 Report on Future Instrumentation for the Anglo-Australian Telescope, March 1978.

4 Although the Board is a bi-national authority, for the purposes of Australian budgetary processes, it is treated like any other governmental body.

5 Report of the Science and Engineering Research Council for the year 1984–85, p.5.

14   Towards the next century

The political winds which rose and abated at various times in Australia and Britain from the late 1970s were a real test of the strength of the Anglo-Australian Telescope Agreement. Throughout these years the best defence the Observatory had against threats to its budget was the bi-national nature of the Board. When the Governments were urging spending restraint, or when currency fluctuations made the AAT seem unusually expensive, the Board members from each national community were able to exert strong influence on their opposite numbers to avoid or to reduce the full effects of financial constraint on the Board's operation. This worked well when the pressure was unilateral, but, even when both Governments were experiencing difficulties, the Board has been able to preserve the financial integrity of the Observatory. This has no doubt been helped by the manifest and continued success of the telescope.

While there has always been unanimity among Board members and within the scientific community about preserving the financial stability of the telescope, there have been some areas of difference. The AAT Agreement is between only two Governments. At one extreme, the British Government agency, the Science and Engineering Research Council (SERC), has regarded it as an international agreement similar to the those between Britain and its European neighbours in such large multi-national projects as CERN, the European Space Agency and the observatory on La Palma. However, the Australian Government has focussed its attention less on the Agreement and more on the Act of the Commonwealth Parliament, without which the Board would have no legal standing in Australia. These differences in perception have given rise to debates about whether the Board and the Observatory should be answerable to the Australian Government in matters of domestic administrative law.

When the AAT project was in its planning phase the two communities had different perceptions of the role the telescope would play. The Australians saw it as a research tool complementing the work in other observatories or the universities. In contrast, for the British it was their first opportunity of regular, guaranteed access to a large, modern, world-class instrument. It held the prospect of seducing back to the British scene some of those scientists who had been attracted elsewhere by the availability of time on telescopes which were suddenly no longer the best available. The AAT became the focus of British optical astronomy.

With the restructuring of the administration of science in Britain under the Science Research Council in 1965, there was an improved system for establishing formal links with the widely representative scientific community. There is still no equivalent formal system in Australia. As a result, the British Board members were in a good position to speak as one voice for their community. For example, they knew what instrumentation their scientists wanted for the AAT, whereas the Australians tended to see special instruments as the responsibility of individuals to develop for themselves. However, once it was operational the Observatory was especially well placed to draw upon the large reservoir of ideas available from the large British scientific community. This has been particularly evident in the provision of instruments and detectors for the telescope.

14.1 The success of the AAT

The success of projects like the AAT, involving substantial initial expenditure, joint planning, on-going financial support and development, rests on a genuine community of interest. With hindsight the AAT Agreement could have been better drafted to spell out clearly the role of a director, or an alternative if the telescope were to have no director of its own. This would have avoided the long, drawn-out debate with the Australian National University about the management of the AAT in the early 1970s. Nevertheless, the AAT continues to be an example of Anglo-Australian scientific collaboration at its very best.

Despite the differences in perception of the Board's role under the AAT Agreement by the two Governments, and the deficiencies inevitable in any agreement, nobody disputes the outstanding practical success of the AAT. Indeed, users of the telescope have said that the AAT is not only the best but the best run of the world's big telescopes; the qualities of design and construction which established its initial reputation have been matched by equally capable and enterprising management. The telescope was accepted almost at once as a technological *tour de force*. The optics set new standards for their time, though the diffraction-limited performance of the AAT primary mirror has since been matched by that of other large mirrors. Equally important was the massive polar axis, which is still unchallenged as a piece of superbly designed and executed engineering, and is at the heart of the stability and precision of the telescope. But what set the AAT apart from other telescopes was the elegance and power of the computer control system, the unprecedented accuracy of its setting and tracking, the facility with which it could move from star to star, and the flexibility that could accommodate important observer aids like beam-switching and automatic focussing. Finally, the revolutionary Image Dissector Scanner made its own distinctive contribution to the initial reception of the telescope.

A decade or more after this initial success the reputation of the AAT stands as high as ever with the emphasis now on efficient management, excellent instrumentation and high productivity. The impact of the telescope on astronomy is second to none. The reasons for the AAT's continuing success are more complex now than those responsible for its initial impact, and we illustrate this by drawing on parallel discussions by Peter Gillingham[1] and Paul Murdin.[2]

First, throughout its existence the AAT has been supported by a powerful Board which has always included influential public servants and distinguished and far-sighted scientists. Its decisions in the planning phase were of fundamental importance, and none more so than that of encouraging the engineers to aim for the highest achievable precision, even when this took them to the limits of engineering technology. Two difficult early decisions which turned out well were the choice of Epping as the site for the laboratory, as opposed to Canberra or Coonabarabran, and the establishing of what is in effect a complementary base at Coonabarabran. The grounds of the CSIRO Radiophysics Division have proved an appropriate, congenial and pleasant base for the Director and part of the staff. Paul Murdin thought that the division into dual sites, which could well have been a recipe for disaster, in practice has had the opposite effect. The interests of the highly competent staff at Coonabarabran, needed in any case to maintain such a sophisticated telescope, are kept fully engaged by on-site developments such as the Gillingham coudé spectrograph (built to observe supernova 1987A), and a continual inflow of new equipment, new software, new ideas and new observers. Moreover, Coonabarabran is at the cutting-edge of the whole organisation; it is where the science is done, where the press and television crews go. The support staff see new results appear on the video screens as soon as the astronomers do, and know, if only from the excitement in the control room, when something big has turned up: heady stuff, which keeps morale high.

In Epping, the stimuli are of a different kind. As in any laboratory, there are numerous technical challenges to be tackled. What is surprising, in view of the relatively small staff, is the range of activities undertaken and the quality and variety of the achievements. There have been notable advances in the fields of electronics, engineering, instrument design and construction, data reduction, software design, optics and optical fibre fabrication and photography. At both ends of the AAT operation the day-to-day problems which running a complex undertaking poses are approached with an obvious willingness to overcome them, often more apparent to visitors than to the staff themselves. It is that spirit which has played an important part in the success of the whole enterprise. Sir Harrie Massey was keen to maintain close contact between the staff and the users, and the association of a scientific symposium twice a year with each Board meeting helps to sustain this spirit in both communities.

Another essential element was the Board's success in securing adequate funds for continued instrument development at a time when scientific funding was becoming increasingly difficult. About a quarter of the budget has consistently gone into the purchase or development of instrumentation, including hardware for data reduction. We have already discussed the instrumentation programme, but we point out again that it is the quality of the instruments which is the observers' chief concern. This applies not only to the instruments themselves, but to the detectors, the possibilities

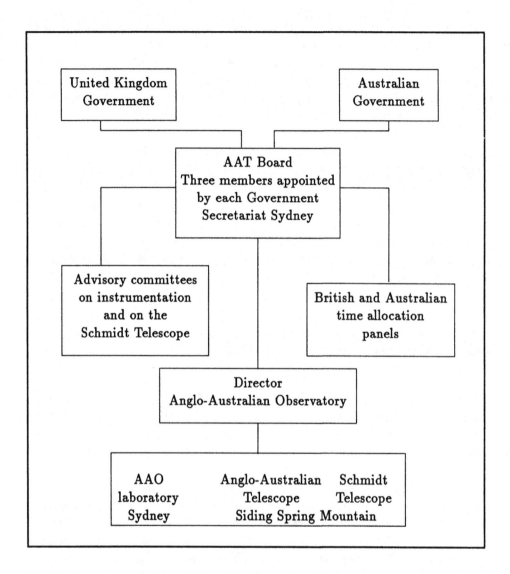

Figure 14.1   Organisation of the Anglo-Australian Observatory in 1989

they create for interactive control, and to the software which can make a system accessible even to a first-time user.

It was realised early that if astronomers, especially visiting astronomers, were to make the best use of the unusually versatile suite of instruments and the opportunities offered by the AAT, they would need a much higher level of support than had been customary at observatories until that time. The quality of the AAT support staff is quite outstanding and has been very important in its success. Here we cannot do better than quote Peter Gillingham:

A key element in the success of the AAT has been the high level of support given observers. Since anyone may apply for time, and a run of more than three nights for the one group is unusual, the users in any year are numerous. With the variety and sophistication of the instruments, virtually all of which are available to all applicants, efficient operation depends critically on supporting the visitors before their run (while they plan their observations and set up their instrument), during the observing, and after their nights (to reduce data).

Each observing team is notified, when its time is allocated, which AAO staff officer is to support it. This support astronomer is the contact for advice before the observer comes to the AAT. Then, unless the observer is fully confident he or she can handle the observations without such help, the support astronomer travels to the telescope to be with the observer during setting up and observing.

The four Night Assistants at the AAT are rostered on seven nights at a time. Between these weeks each works at daytime duties, not as a fill-in, but as a full member of the technical staff. Four electronics technicians work a roster of evening shifts, a week at a time, starting at 2 pm and ending at about 10:30 pm, so that they overlap with the day staff and are present for the first few hours of observing, when faults are most likely. This evening technician is on call after he ends his shift and over the weekend. Confidence that most faults can be cured on the night with minimal time lost has characterised the operation. Over the last five years the proportion of time lost due to AAT equipment failure has averaged 2.3 per cent.

Minimising the distinction between day and night staff has helped a good deal in avoiding discord and it naturally makes the Night Assistants more knowledgeable about the technicalities of instruments and the day staff more appreciative of the problems facing astronomers at night.[3]

Although the AAT was envisaged as a facility for astronomers from Britain and Australia, its outstanding performance and results quickly encouraged scientific collaborations beyond these national boundaries. British and Australian observers have undertaken valuable programmes with foreign astronomers from more than a dozen other countries. Figure 14.2 shows the allocation of AAT observing time by the two national time assignment panels over the period September 1980 to August 1989. The time awarded to foreign institutions has steadily grown and they rank as a group with the three major British user groups.

Another advantage of the AAT, one not shared by observers using its counterparts in Chile, is ready access to the big Australian radio observatories. This has been

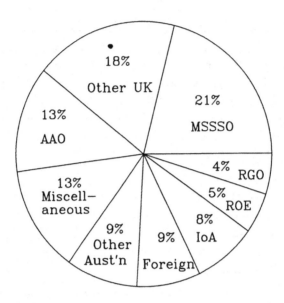

Figure 14.2   Time allocation by PATT and ATAC 1980–1989

especially important because the Parkes 64-metre dish and the Molonglo telescope are between them responsible for almost all the data on the southern radio sources. Other installations, such as the NASA Deep Space Station near Canberra with its 70- and 34-metre radio telescopes, are also occasionally part of the Australian radio astronomy network. The discovery of the optical counterpart of the Vela pulsar would have been impossible without the intimate co-operation of the radio observatory at Parkes, and the University of Sydney's radio arrays at Molonglo and Fleurs. Even more important was the collaboration between the AAT and the UK Schmidt Telescope. Plates from the Schmidt supplied a steady stream of information to the AAT, on optical identifications for X-ray and radio sources, on new and peculiar galaxies, emission nebulae, planetary nebulae, and an endless variety of phenomena and objects. Equipped with its new objective prisms, the Schmidt became the most efficient instrument in existence for finding new and ever more distant quasars, and enabled the AAT to be the leader in this particular field.

Important, too, is Siding Spring Mountain as an astronomical site. We have to balance the claims of the better climate in Chile and on La Palma against the more sophisticated level of operations possible in Australia, and the agreeable working environment on Siding Spring and in nearby Coonabarabran, especially against that at Mauna Kea, on Hawaii. Very important is the technical support which is particularly strong in Sydney relative to the corresponding centres nearest the telescopes in Chile, and on La Palma and Mauna Kea. Astronomers can observe about 80 per cent of the time in Chile, as opposed to 60 per cent on Siding Spring because of weather, and the average seeing may well be better in Chile, though Siding Spring gets its share of sub-arc second seeing. But world-class astronomy is performed in Australia, as evidenced by the production record of the UK Schmidt Telescope, and the quality of its SERC(J) atlas; and by the fact that over the last decade about one hundred papers have been produced each year from data gathered with the AAT. It seems a moot point as to whether a more remote site, which may be better astronomically but more isolated and more inhospitable in human terms, would produce more and better astronomy.

14.2 The influence of the AAT

There can be no denying the impact of the AAT. Its technical influence has been widespread, and together with the UK Schmidt it has not only been the principal force in a great resurgence of British optical astronomy, but has come to dominate Australian optical astronomy in much the same way. Had it not been for the AAT, and the depth of engineering talent in both countries which its success revealed, it is doubtful whether either the 4.2-metre William Herschel or the 2.3-metre ANU telescopes would ever have been built.

The AAT was not only a great success technically, but it changed the style in which astronomy was pursued. The new control systems of the telescope itself and of the auxiliary instruments made observing so much easier that suddenly many of the traditional observers' skills had become redundant, and the telescope was available to a much wider circle of users. In addition, the decision to designate it as a national facility, which was a new departure in Australia if not in Britain, made it effectively open to all comers. Anyone with a reasonable research proposal could apply for observing time, regardless of his or her professional affiliation, and time in both countries was allocated by independent committees of astronomers with broadly based and frequently changing membership. These are the UK Panel for the Allocation of Telescope Time and the Australian Time Assignment Committee. In both countries it is possible for small university departments to carry out front-line research with the use of the best equipment available without having a telescope of their own.

In Britain this principle of common access has been extended to facilities for data reduction. For some years the SERC establishments and the principal university astronomy departments have been connected to a common computer network called Starlink, through which they have access to the most up-to-date and sophisticated computer programs. This is a vital adjunct to contemporary astronomy, be it theoretical or observational. The AAO is linked to the Starlink system via satellite and cable with direct access to the British institutions on the Starlink network.

On the more technical side, the impact of the AAT control system was immediate, and almost all large telescopes built since have incorporated similar systems. It was, of course, inevitable that computer control would become important. The particular contribution of the AAT was to demonstrate how necessary it was to integrate the control system with the telescope from the outset, so that full allowance could be made for its impact on the mechanical and other components of the design. In the case of the AAT, it was to accommodate the requirements of the computer that the polar axis was redesigned, spur gears were adopted for the drive, and the particular configuration of absolute and incremental encoders was chosen.

Another contribution of the AAT was that it made the alt-azimuth mounting for optical telescopes a practical, indeed inevitable, proposition, although not itself an alt-azimuth telescope. This is not to deny the claims of the Russian 6-metre telescope which was clearly the first in this field; but communications with the USSR were poor at the time, there were initial problems, and some years elapsed before the telescope began to make an impact. The problem with an alt-azimuth was not so much the provision of continually varying drive speeds in each co-ordinate, as the need to realise them to the fractional arc second accuracy demanded by the optical astronomers. Straightforward extensions of an AAT-like control system were all that was needed to meet this requirement. The mechanical and other advantages of the alt-azimuth mounting are well-known. One is its saving in weight. Though its mirror is slightly larger than that of the AAT, the William Herschel Telescope with its alt-azimuth mounting weighs only two-thirds as much. Another advantage, not often mentioned, is that it can bypass one of the most serious operational problems of an equatorial telescope. We refer to instrument changes at the Cassegrain focus. Besides awkward mechanical adjustments and a myriad of electronic connections, instrument changes now often involve such complications as coolants for electromagnetic coils and cryogenics for detectors, and become steadily more difficult and time consuming to implement. An alt-azimuth telescope, with two Nasmyth foci as well as a normal Cassegrain, is much more flexible and accessible in this respect.   •

14.3 The AAT and the new optical telescopes

We have described how for several decades there was no great incentive to build a bigger telescope than the Hale 5-metre. Especially in the 1960s and 1970s it was more profitable to utilise the steadily increasing efficiency of the new light detectors than to push up the size of the primary mirror. But in the 1980s when the quantum efficiencies of charge-coupled devices began consistently to exceed 50 per cent it was clear that that road was coming to an end, and thoughts turned again towards building bigger telescopes. A scaled-up Anglo-Australian or Kitt Peak design for, say, an 8-metre telescope, was ruled out on grounds of weight and cost. The mirror would be excessively heavy and costly, and the mounting, even an alt-azimuth mounting, prohibitively massive. The time had come for a complete break with the ponderous 'battleship' style exemplified by the Hale and later telescopes. In any new telescope the mirror had to be drastically reduced in weight, and the building in size. If this meant an ultra-short focal length it also meant a stiffer tube and less wind loading. The latter caused increasing concern with the trend towards more open domes and the maximum feasible exposure of the mounting to ambient conditions adopted to avoid dome seeing problems. The heavy battleships were becoming lightweight frigates – fast, modern and efficient.

Without going into detail inappropriate in this book, there have been four lines of attack in the development of the next generation of optical telescopes. All involve alt-azimuth mountings which seem now to be taken for granted. There is a general consensus that eight metres is the practical limit for a single mirror; bigger mirrors are too expensive to cast, too flexible to figure and mount, and too difficult to transport especially to a remote site on another continent. In addition, the problems of building vacuum systems for aluminising them are overwhelming. It follows that telescopes with apertures greater than eight metres must be built with multiple objectives. These may consist of an array of identical and identically mounted telescopes, the beams of which are brought to a common focus either by trains of smaller mirrors or by optical fibres. Alternatively, the mirrors may be tesselated, formed of a number of segments figured and fitted together so that they jointly constitute a continuous optical surface, such as a paraboloid.

Both ideas are being pursued. The first is exemplified by the Multiple Mirror Telescope located on Mount Hopkins, Arizona. It has six identical mirrors, each 1.83 metres in diameter, mounted in hexagon fashion on a common altitude axis. Their beams are united at a common focus by trains of secondary mirrors to produce the equivalent of a single telescope of 4.4 metres aperture. This remarkable telescope broke new ground, and while there have been problems combining the output of the six component mirrors consistently, the system has worked successfully since its commissioning in 1979. To provide greater light gathering power there are now plans to replace the six mirrors with a single monolithic mirror of 6.5 metres diameter.

A second approach to new telescope design is seen in the Keck Telescope which will have a 10-metre aperture. It is being built jointly by the California Institute of Technology and the University of California, and will be located on Mauna Kea. Its mirror will be a mosaic of 36 hexagons, each 1.8 metres across, fitted together and arranged in circles of 6, 12 and 18 units. The positions of the hexagons relative to one another will be measured with sensors, and they will be supported on systems of actuators which will push against them until they are correctly adjusted. They then will form a single continuous surface of the required hyperboloidal shape; the precision required for this operation, of course, is extreme. The hexagons will have been previously figured to the correct non-spherical shapes. As they are all to be mounted off-axis, and the mirror will be very fast, this too is no easy task.

A third idea being actively pursued by the European Southern Observatory and by the Japanese is to use a very thin mirror and to control its shape by an active support system. This system is similar in principle to that on the Keck Telescope, but there are substantial differences in detail. A prototype of this, the ESO New Technology Telescope (NTT), is currently being commissioned in Chile. It has a solid meniscus mirror made of Zerodur, 3.5 metres diameter and 22 cm thick, and is carried on an 'active' support system of 75 actuators and three fixed points. It too is on an alt-azimuth mounting, with a total weight in its case of 120 tonnes. There is a sketch of the mounting in Figure 7.1(f). To adjust the mirror the telescope is pointed at a bright star, and the image, or reflected wavefront, is examined by a test device rather like an on-line electronic Hartmann test (Shack device), which measures and analyses the data in real time. It tells the actuators how to correct the mirror's figure, and they push against the mirror and adjust its shape accordingly. The process can be repeated as often as desired. The NTT is intended as a pilot design for an array of four 8-metre telescopes which will be carried on a common mounting. It will then be possible to combine at a common focus the light collected by each telescope using optical fibres. There are various American plans for large telescopes, and the Japanese also hope to build a 7.5-metre instrument with a mirror 20 cm thick supported on 400 active points.

Active optics is high technology, but while it has its attractions it is expensive and may have other problems. Roger Angel of the University of Arizona works towards circumventing these by casting mirrors in which the body is not solid but made of thin-walled hexagonal cells, honeycomb style. His technique derives from that of the 5-metre Hale mirror, but the walls of his cells are much thinner, and he uses a commercial crown glass which is cheap, has a relatively low melting-point, and flows and pours easily. Although it expands and contracts on heating much more than other vitreous substances discussed in this book, he expects that because of their cellular construction his mirrors can be maintained at constant temperature by forced air ventilation, aided by holes in the walls of the hexagonal cells. Since the mirror's upper surface is curved, the furnace is spun during cooling. Then the softened glass will assume the shape of a paraboloid, and the curvature will depend on the rate of spinning. This achieves great economies in materials and manufacture. Angel has supplied one such 3.5-metre blank for the telescope made for the Astrophysical Research Consortium, a group of five American universities. The mirror is to work at f/1.75 (focal length 6.1 metres), and will weigh only a quarter as much as an equivalent solid disc. The telescope will have an alt-azimuth mounting and a total weight of 45 tonnes, half that of the AAT horseshoe. As with ESO's NTT, the first reports on its performance are eagerly awaited.

14.4 The AAT and the future

When these and other giant telescopes come to fruition over the next decade or two, will the AAT be outclassed by a new generation of telescopes which incorporate at least some of the ideas it pioneered? History does not support this view. As time passes there will be changes in the role of the AAT, but there will always be a demand for a telescope which can point and drive accurately, and which has good optics. If in addition, its instruments are kept up to date it will be in heavy demand. Telescopes are in fact remarkably durable, certainly as compared with the high energy accelerators of modern particle physics, instruments with which they otherwise have features in common. Given a sound engineering design and good optics, they lend themselves well to periodic up-dating, and can usually accommodate new instruments without undue difficulty. The Great Melbourne Telescope (1869) was still working in the 1970s, and was retired then because of the heavy demands on the Mount Stromlo engineering services made by the newly authorised ANU 2.3-metre telescope. The 36-inch Crossley (1879) is still in use at the Lick Observatory; in the hands of the highly competent Lick staff it held its own remarkably well for a time with the larger Mount Wilson telescopes. It is true that the Mount Wilson 60- and 100-inch are no longer used, but their retirement was hastened by the tremendous increase in the brightness of the Los Angeles night sky immediately to their west.

Another factor which increasingly effects telescope usage is the availability of observing time. Until fairly recently the few large telescopes in existence were the property either of individual universities or of quasi-private bodies like the Carnegie Institution of Washington. First call on their time went to their resident astronomers, who with no one to compete with but one another, could plan long-term programmes which might have little to show in the short term, but were nevertheless of the highest importance. Classic examples, familiar to all astronomers, are Shapley's investigation of the globular clusters, which revealed the true nature of our own galaxy and of the sun's place in it; and the discovery by Hubble and others of the extragalactic Universe and the fact and manner of its expansion. Both investigations occupied large fractions of the time of the Mount Wilson telescopes.

Almost all the current large telescopes listed in Appendix 9 are either owned by consortia or are national facilities, and therefore are open to a much wider circle of potential users than was formerly the case. Without exception they are over-subscribed by factors of three to five, and only in extreme circumstances can any one group expect an extended run or as many as a dozen nights a year. Emphasis therefore goes very much to short-term programmes in which one or two nights observing will yield a quick return, and so provide a basis for further applications for observing time. Such a system does not lend itself to experiment; observers tend to play safe with the biggest telescopes, reserving for those further down the scale trials of new instruments, and the risky but often excitingly productive innovations. In the same way the building of the ANU's 2.3-metre telescope was not at all discouraged by the presence of the AAT, nor was the move of the Isaac Newton Telescope to La Palma by the plans for the William Herschel Telescope.

Let us hazard a guess into a future which includes a southern hemisphere telescope of eight metres aperture embodying all the latest ideas, and may even be located in Australia. This telescope would take over much of the spectroscopy of the really faint objects, such as remote quasars, and the globular cluster companions of distant galaxies. Other things being equal, the ability of this telescope to reach the same accuracy four times as quickly as the AAT would be decisive, and the fainter the object the more important it would become. For critical definition at faint limits the 8-metre might again have an advantage, less certain because it would depend on external factors like site selection; more importantly, it then presumably would be competing with the Space Telescope. For brighter quasars the AAT might retain much of its present ascendency; it will be a long time before anything in the south will approach the close-knit combination of the AAT, the UK Schmidt Telescope and the various radio telescopes, now reinforced by the Australia Telescope. For multi-object spectra the AAT promises to hold a particular advantage in that its optical system suitably enhanced will be able to produce fields of view much wider than those available on an 8-metre class telescope. Finally, it would be freer to pursue long-term programmes requiring larger allocations of time, and this could become important. Such programmes have been running for some time on the ANU's 2.3-metre and smaller telescopes. The future of the AAT is assured for many years, and there is ample room for it and for large telescopes.

The AAT Board has always taken a very positive view of proposals for new instrumentation. It has given encouragement to its own Advisory Committee on Instrumentation (ACIAAT) and ensured that a substantial proportion of the AAT budget has been available for the construction of new instruments. We conclude by describing briefly three state-of-the-art instruments actively being considered for the AAT. They will give the flavour of what is being planned to keep the telescope in the forefront of its subject. The first is an infrared camera-spectrometer built around a new array detector, which covers a whole field instead of one tiny element. It promises to increase efficiencies by a factor of ten to a hundred, and will allow pictures to be taken at infrared wavelengths in the same way that the optical pictures are made using CCD detectors.

The second is a successor to the existing multi-object spectrograph which will place 300 or more fibres over a two degree field at the prime focus. This will necessitate the design and manufacture of a large new corrector lens to be mounted on a new top-end on the telescope, an advanced new fibre positioner, and several new spectrographs. The system will greatly increase the power of the AAT in this field, and will lead to results unobtainable with any other telescope, including most of those currently proposed in the 8-metre class.

Finally, there is a proposal to build an auxiliary telescope alongside the AAT, specifically to feed light to the highly successful coudé echelle spectrograph, which

Figure 14.3   A ten-hour exposure of stars moving around the south celestial pole. Four stars in the Southern Cross and the Pointers, α and β Centauri, are marked.

could then be used independently of the main telescope itself. Using modern concepts such as a segmented mirror and optical fibres, such a telescope could be built relatively cheaply with a light-gathering power comparable to that of the AAT. It could thus double the scientific output of the AAT and open up several completely new lines of research.

The contribution of the AAT to astronomy and astrophysics is well established. A rewarding feature of this contribution is the spirit of international collaboration in which it is made, and the enthusiasm of all those who are associated with it. Less well-known is the formal basis for the operation of the Anglo-Australian Observatory: the Anglo-Australian Telescope Agreement. It is the only agreement between the Governments of the United Kingdom and Australia in science, and has proved to be an outstandingly successful example of scientific collaboration between the two countries.

Long before the telescope was invented, the Universe held a fascination for man who has used astronomy to discover more about the Universe and our place in it. With each wave of improved technology we have the means of answering more of the questions which the Universe presents. These answers often have been the inspiration for ever more numerous and more demanding observations. This is the sign of a healthy, indeed flourishing branch of science. Despite the enormous growth of knowledge in recent years, there remain many more astronomical problems to solve. Some can only be undertaken by telescopes bigger than the AAT but most will remain within its grasp for many decades. As long as novel ideas abound, the AAT will remain a world-class instrument. There is no sign at present that the flow of ideas is drying up.

Notes to Chapter 14

1 Gillingham P.R. (1989). Lessons for New Large Telescopes from the AAT, *Japanese National Large Telescope and related engineering development,* T. Kogure & A.T. Tokunaga (ed.) 281

2 Personal communication between Murdin and the authors after the A. Hoag Retirement Conference in Flagstaff, Arizona, 1985

3 Gillingham P.R. (1989). Lessons for New Large Telescopes from the AAT

Appendix 1   AAT Agreement Act 1970

which includes the Anglo-Australian Telescope Agreement

ANGLO-AUSTRALIAN TELESCOPE AGREEMENT ACT 1970

An Act relating to an Agreement between the United Kingdom and the Commonwealth with respect to the Establishment and Operation in Australia of an Optical Telescope

PART I—PRELIMINARY

1. This Act may be cited as the *Anglo-Australian Telescope Agreement Act* 1970.[1]   Short title

2. This Act shall come into operation on a date to be fixed by Proclamation.[1]   Commencement

*   *   *   *   *   *   *   *   Section 3 repealed by No. 216, 1973, s. 3

4. In this Act, unless the contrary intention appears—   Interpretation

"Australian member" means a member, or a temporary member, of the Board appointed by the Governor-General in accordance with section 7 of this Act;

"member" means a member of the Board, and includes a temporary member;

"the Agreement" means the Agreement set out in the Schedule to this Act;

"the Board" means the Anglo-Australian Telescope Board;

s. 4

" the telescope" has the same meaning as in the Agreement.

Approval of Agreement

5. The Agreement a copy of which is set out in the Schedule to this Act is approved.

PART II—ESTABLISHMENT AND FUNCTIONS OF THE TELESCOPE BOARD

Establish-
ment of
Board

6. (1) For the purposes of this Act and of the Agreement, there shall be an Anglo-Australian Telescope Board.

(2) The Board is a body corporate, with perpetual succession and a common seal.

Constitution
of Board

7. (1) The Board shall be constituted as provided in the Agreement, and meetings of the Board shall be conducted in accordance with the Agreement.

(2) The rights and powers of the Government of the Commonwealth under the Agreement to appoint members and temporary members of the Board shall be exercised by the Governor-General.

(3) Subject to the next succeeding sub-section, an Australian member shall be paid by the Commonwealth such remuneration, and holds office on such other terms and conditions, as the Governor-General determines.

(4) The Minister may grant leave of absence to an Australian member on such terms and conditions, including terms and conditions as to remuneration, as the Minister determines.

Functions,
capacities
and powers
of Board

8. The Board has the functions specified in Article 8 of the Agreement, and the capacities and powers specified in paragraph (1) of Article 6 of the Agreement, and shall perform those functions, and exercise those capacities and powers, in accordance with the Agreement.

Resignation
of Australian
members

9. An Australian member may resign his office by writing under his hand addressed to the Governor-General.

Removal of
members

10. (1) The Governor-General may remove an Australian member from office for misbehaviour or physical or mental incapacity.

(2) If an Australian member—

(a) becomes bankrupt, applies to take the benefit of any law for the relief of bankrupt or insolvent debtors, compounds with his creditors or makes an assignment of his remuneration for their benefit;

(b) is absent, except on leave granted by the Minister, from three consecutive meetings of the Board; or

(c) fails to comply with paragraph (4) of Article 7 of the Agreement, the Governor-General shall, by notice published in the *Gazette*, declare that the office of the member is vacant, and thereupon the office shall be deemed to be vacant.

PART III—FINANCE

11. (1) There are payable to the Board such moneys as are appropriated by the Parliament for the purposes of the Board.

Moneys payable to Board

(2) The Minister for Finance may give directions, not inconsistent with the Agreement, as to the amounts in which, and the times at which, moneys referred to in the last preceding sub-section are to be paid to the Board.

Amended by No. 36, 1978, s. 3

12. (1) The Board may open and maintain an account or accounts, in Australia or elsewhere, with an approved bank or approved banks and shall maintain at all times at least one such account.

Bank accounts

(2) The Board shall pay all moneys of the Board, including moneys received under the last preceding section, into an account referred to in this section.

(3) In this section, "approved bank" means the Reserve Bank of Australia or another bank approved by the Treasurer.

13. (1) The moneys of the Board shall be applied only in payment or discharge of expenses, obligations and liabilities of the Board.

Application of moneys

(2) The Board may not expend moneys of the Board otherwise than in accordance with estimates of expenditure prepared by the Board and approved in accordance with the Agreement.

14. (1) The Auditor-General shall inspect and audit the accounts and records of financial transactions of the Board and records relating to assets of, or in the custody of, the Board, and shall forthwith draw the attention of the Minister to any irregularity disclosed by the inspection and audit that is, in the opinion of the Auditor-General, of sufficient importance to justify his so doing.

Audit

(2) The Auditor-General may, at his discretion, dispense with all or any part of the detailed inspection and audit of any accounts or record referred to in the last preceding sub-section.

(3) The Auditor-General shall, at least once in each year, furnish to the Minister a report on the results of the inspection and audit carried out under sub-section (1) of this section.

s. 14

(4) The Auditor-General or an officer authorized by him is entitled at all reasonable times to full and free access to all accounts, records, documents and papers of the Board relating directly or indirectly to the receipt or payment of moneys by the Board or to the acquisition, receipt, custody or disposal of assets by the Board.

(5) The Auditor-General or an officer authorized by him may make copies of, or take extracts from, any such accounts, records, documents or papers.

(6) The Auditor-General or an officer authorized by him may require a member or employee of the Board to furnish him with such information in the possession of the member or employee or to which the member or employee has access as the Auditor-General or the authorized officer considers necessary for the purposes of the functions of the Auditor-General under this Act, and the member or employee shall comply with the requirement.

Exemption from taxation

Amended by No. 216, 1973, s. 3

15. The Board is not subject to taxation under any law of the Commonwealth or of a State or Territory.

PART IV—MISCELLANEOUS

Compensation (Commonwealth Employees) Act to apply

Sub-section (1) amended by No. 51, 1971, s. 3

16. (1) The *Compensation (Commonwealth Employees) Act* 1971[2] applies to an Australian member as if he were an employee within the meaning of that Act, and for that purpose an Australian member shall be deemed to have entered into a contract of service with the Commonwealth for the performance of his duties as a member.

Amended by No. 51, 1971, s. 3

(2) The *Compensation ('Commonwealth Employees) Act* 1971[2] applies to employees of the Board as if they were employees within the meaning of that Act and as if references in that Act to the Commonwealth were references to the Board.

Retention of Public Service rights for eligible employees

Amended by No. 216, 1973, s. 3

17. Where a person employed by the Board was, immediately before becoming so employed, an officer of the Public Service of the Commonwealth—

(a) he retains his existing and accruing rights; and

(b) for the purpose of determining those rights, his service under this Act shall be taken into account as if it were service in the Public Service of the Commonwealth.

Evidence of appointment of United Kingdom member

18. In proceedings in any Court, a certificate under the hand of the Minister certifying that a person specified in the certificate is, or was at a specified date, a member or temporary member of the Board duly appointed by the Government of the United Kingdom, or certifying as to any matter concerning the duration of such an appointment or the terms and conditions applicable to such an appointment, is evidence of the matter so certified.

19. (1) The financial statements to be furnished by the Board in ac- Reports cordance with Article 8 of the Agreement shall be in such form as the parties to the Agreement approve, and the Minister shall cause a copy of the financial statements so furnished in respect of any financial year to be submitted to the Auditor-General, who shall report to the Minister—

(a)   whether the statements are based on proper accounts and records;

(b)   whether the statements are in agreement with the accounts and records;

(c)   whether the receipt and expenditure of moneys, and the acquisition and disposal of assets, by the Board during the year have been in accordance with this Act and the Agreement; and

(d)   as to such other matters arising out of the statements as the Auditor-General considers should be reported to the Minister.

(2) The Minister shall cause a copy of each report on the operations of the Board furnished in accordance with Article 8 of the Agreement and of the financial statements furnished with the report, together with the report of the Auditor-General on those financial statements, to be laid before each House of the Parliament within fifteen sitting days of that House after receipt by the Minister of the report of the Auditor-General.

(3) The approval of the Government of the Commonwealth to the Amended by form of the financial statements to be furnished by the Board shall be s. 3 No. 36, 1978, given by the Minister for Finance.

THE SCHEDULE   Section 5

AGREEMENT

BETWEEN THE GOVERNMENT OF THE COMMONWEALTH OF AUSTRALIA AND THE GOVERNMENT OF THE UNITED KINGDOM OF GREAT BRITAIN AND NORTHERN IRELAND TO PROVIDE FOR THE ESTABLISHMENT AND OPERATION OF A LARGE OPTICAL TELESCOPE

The Government of the Commonwealth of Australia and the Government of the United Kingdom of Great Britain and Northern Ireland;

Desiring to provide for the establishment and operation in Australia as a joint enterprise of a large optical telescope that will enable Australian and United Kingdom astronomers to undertake astronomical observations to the advancement of scientific knowledge;

THE SCHEDULE—continued

Have agreed as follows:

Article 1

(1) In the Agreement, unless the context otherwise requires:

(a) "the Commonwealth Government" means the Government of the Commonwealth of Australia;

(b) "the United Kingdom Government" means the Government of the United Kingdom of Great Britain and Northern Ireland;

(c) "the telescope" means the telescope constructed in accordance with this Agreement and, where the context so admits, includes ancillary equipment, facilities and services essential to the effective operation of the telescope;

(d) "the Telescope Board" means the Anglo-Australian Telescope Board established in accordance with this Agreement;

(e) "the University" means the Australian National University constituted under the Australian National University Act 1946-1967 of the Commonwealth of Australia.

(2) For the purposes of this Agreement each Contracting Party shall act through a designated agency. These two agencies shall be responsible jointly for implementing the Agreement. For the purposes of making such agreements, determining such matters or receiving such consents, approvals or communications as may be required for the operation of the Agreement, each agency shall be represented by a person nominated by that agency.

ESTABLISHMENT OF THE TELESCOPE
Article 2

(1) The Contracting Parties shall cause to be manufactured, constructed, operated and maintained, by the Telescope Board, an optical telescope and associated facilities and services. The telescope shall have a nominal aperture of 150 inches.

(2) The specifications for the telescope shall be based on the design adopted by the Association of Universities for Research in Astronomy for a similiar optical telescope to be located at Kitt Peak, Arizona in the United States of America, but shall incorporate such modifications to that design as are agreed by the Telescope Board to be necessary in the light of the desired uses of the telescope.

(3) The telescope shall be the property of the Telescope Board.

Article 3

(1) World-wide tenders shall be called for the optics, mounting, telescope tube, building, dome and other major components of the telescope, unless it is agreed by the Telescope Board that tenders should be invited on a selective basis from firms known to be competent. The Telescope Board shall decide which components are to be regarded as major components and shall approve in advance of calls for tenders for major components, the specifications and list of tenderers.

(2) If the members of the Telescope Board do not agree unanimously on the placement of a particular contract the matter shall be referred by the Telescope Board to the Contracting Parties for determination. The placement by the Telescope Board of contracts for major components shall be approved by the Contracting Parties.

Article 4

(1) The Commonwealth Government shall arrange with the University for the use by the Telescope Board of a site for the telescope in the area that is vested in or under the control of the University at Siding Spring Mountain in the State of New South Wales. The terms and conditions of such use shall be as agreed upon between the Telescope Board and the University.

(2) So far as practicable and subject to satisfactory arrangements being made between the Telescope Board and the University, use should be made of supporting facilities in existence or to be provided by the University at Siding Spring Mountain and at Mount Stromlo. This does not however preclude the use of supporting facilities elsewhere.

THE SCHEDULE—continued

(3) The arrangements for the provision by the University of facilities and services for the purposes of construction, operation and maintenance of the telescope shall be such as are agreed upon between the Telescope Board and the University, and the Commonwealth Government shall accord its good offices as appropriate in the negotiations and the putting into effect of these arrangements.

OPERATION AND USE OF THE TELESCOPE

Article 5

(1) Observing time on the telescope and use of associated joint facilities and services shall be shared equally between the Contracting Parties and it shall be the responsibility of the Telescope Board to make arrangements consistent with this principle in its determination of the use of the telescope and associated facilities and services.

(2) The rules to be made by the Telescope Board shall include rules governing the detailed arrangements for manintenance, operation and use of the telescope.

(3) Either Contracting Party may permit other than United Kingdom or Australian astronomers to use the telescope during the time to which that Party is entitled to use the telescope in accordance with this Article.

THE TELESCOPE BOARD

Article 6

(1) The Telescope Board to be incorporated under an enactment of the Parliament of the Commonwealth of Australia shall be a body corporate with perpetual succession and a Common Seal and shall have such capacities and powers as are necessary and incidental to the performance of its functions under this Agreement including, without affecting the generality of the foregoing capacities and powers:

(a) to acquire, hold and dispose of real and personal property;

(b) to enter into contracts including contracts for the performance of works and contracts of service and for services;

(c) to employ persons;

(d) to sue and be sued;

(e) to receive gifts;

(f) to do anything incidental to any of its powers.

(2) The Telescope Board shall be constituted of as many members as are agreed by the Contracting Parties always provided each Party is equally represented. Unless otherwise agreed, each Party shall appoint for the time being three members on terms and conditions determined by the Party making the appointment.

(3) Each Contracting Party may appoint a person to be a temporary member, whenever a member previously appointed by that Party is absent or unavailable. During the term of his appointment a temporary member shall have and may exercise all the functions of a member of the Telescope Board.

Article 7

(1) The Telescope Board shall make arrangements for two of its members to act as Chairman and Deputy Chairman respectively. The periods of tenure and the functions of these offices shall be as determined by the Telescope Board. The functions of the Chairman, or, in his absence or inability otherwise to perform his functions, of the Deputy Chairman, shall include convening and presiding at meetings of the Telescope Board and carrying out such executive responsibilities and functions as the Telescope Board considers to be appropriate subject to and pending any directions or decisions of the Telescope Board.

(2) In the event of the absence of the Chairman and the Deputy Chairman from any meeting of the Telescope Board, the members present shall appoint one of their members to preside at that meeting.

(3) Subject to this Agreement, the Telescope Board may make rules with respect to the order and conduct of the business at its meetings.

(4) A member of the Telescope board who is directly or indirectly interested in a contract made or proposed to be made by the Telescope Board, otherwise than as a member and in common with others of an incorporated company consisting of not less than twenty-five persons,

THE SCHEDULE—continued

shall as soon as possible disclose the nature of his interest at a meeting of the Telescope Board. Thereafter, unless the Telescope Board gives permission, the member shall take no part in any decision of the Telescope Board in respect of that contract.

(5) Four members of the Telescope Board, being two members appointed by each Party, shall constitute a quorum for the transaction of the business of the Telescope Board.

Article 8

(1) The functions of the Telescope Board shall be to do or arrange or cause to be done, subject to and in accordance with Article 2 of this Agreement, such acts, things and matters as shall provide for or contribute to the manufacture, construction, operation and management of the telescope and, without prejudice to the generality of those functions, specific functions shall be:

(a) to approve the final specifications for the telescope, ancillary equipment and buildings and other works;

(b) to arrange for and supervise construction;

(c) to employ persons under such conditions as shall be approved by the Telescope Board in accordance with Articles 17 and 18;

(d) to appoint agents, consultants and advisers;

(e) to enter into arrangements with the University;

(f) to receive, expend and account for funds;

(g) to arrange for, direct and control the operation of the telescope;

(h) to arrange, with the approval of the Contracting Parties, for such modifications to the telescope subsequent to its initially being brought into use as may be necessary from time to time for its more efficient operation;

(i) to furnish to the Contracting Parties as soon as practicable after each 30th day of June, a report on the operations during the year ending on that date together with financial statements in respect of that year and statements of estimated expenditure in future years in accordance with Article 16.

(2) The first report to be furnished by the Telescope Board in accordance with subparagraph (i) above shall relate to the period ending on the 30th day of June, 1970.

Article 9

(1) The Telescope Board may appoint advisory or executive committees to act in connection with the performance of its functions and may determine and direct the functions and procedures of committees so appointed.

(2) An advisory or executive committee may, if the Telescope Board considers fit, consist of or include persons who are not members of that Board.

Article 10

The principal office of the Telescope Board shall be located in the Australian Capital Territory.

FINANCIAL ARRANGEMENTS

Article 11

(1) The costs of manufacturing and constructing the telescope shall be borne by the Contracting Parties in equal shares.

(2) Except as otherwise agreed by the Contracting Parties in particular cases, all costs for operation, use and maintenance of the telescope shall be borne by the Contracting Parties in equal shares.

(3) The Telescope Board shall determine the financial and accounting arrangements for the manufacture, construction, operation, use and maintenance of the telescope, in accordance with Article 5 and paragraphs (1) and (2) of this Article.

Article 12

The costs referred to in Article 11 shall comprise all costs and expenses incurred by the Telescope Board in connection with the performance of its functions and shall include such commitments made after the 31st day of August, 1967 before the formation of the Telescope Board,

THE SCHEDULE—continued

expenditure related thereto and any other expenditure which would have fallen to the Telescope Board had the Telescope Board existed when it was incurred.

Article 13

The costs referred to in Article 11 of this Agreement shall not, unless the Telescope Board otherwise so approves, include expenses of the Contracting Parties except as provided for in Article 12.

Article 14

The Contracting Parties shall make payments to the Telescope Board from time to time in such amounts as are required to enable it to meet the expenditures necessary to the proper carrying out of its functions under this Agreement and within the limits of estimates approved in accordance with Article 16.

Article 15

(1) The Telescope Board shall:

(a)   cause to be kept proper accounts and records of its transactions and affairs;

(b)   ensure that all payments are properly authorized and correctly made;

(c)   ensure that adequate control is maintained over its assets and over the incurring of liabilities.

(2) The accounts, records and financial transactions of the Telescope Board shall be audited from time to time by the Auditor-General for the Commonwealth of Australia. Copies of the reports relative to the audit of the accounts, records and financial transactions shall be supplied to both Contracting Parties.

Article 16

The Telescope Board shall prepare and submit to the Contracting Parties for their approval detailed annual estimates of receipts and expenditure covering accounting periods ending on the 30th day of June in each year together with outline estimates of expenditure for the following five years. The Telescope Board shall prepare and submit such other detailed estimates as may be required by either Party. The estimates shall be in such form as to permit the extraction of information covering annual periods ending on the 31st day of March in each year.

EMPLOYMENT OF STAFF

Article 17

The Contracting Parties and the Telescope Board shall take such action as is appropriate and competent to be taken on their respective parts to provide for the preservation and retention by persons who are employed by the Telescope Board of the existing and accruing employment rights of those persons.

Article 18

For the purpose of the employment of persons, the Telescope Board shall normally follow terms and conditions used by the Commonwealth Public Service, but these terms and conditions may be varied by the Telescope Board when it considers this is necessary and in particular when it is necessary to provide, in accordance with Article 17, for the preservation and retention of the existing and accruing employment rights of those persons normally resident in the United Kingdom.

GENERAL

Article 19

The Commonwealth Government shall take all steps that are appropriate and practicable for it to take to ensure that development or activities do not take place or are not carried out in the area of Siding Spring Mountain that would cause interference with the effective use of the Telescope.

Article 20

(1) The Commonwealth Government shall facilitate entry into, or exit from, Australia free from customs or other duties of goods required in connection with the manufacture, construction, operation, use or maintenance of the telescope and its ancillary equipment which, at the time of entry for home consumption, are the property of the Telescope Board and are not

THE SCHEDULE—continued

goods of a kind, which if produced or manufactured in Australia, would be subject to excise duties.

(2) Exemption from sales tax shall be allowed by the Commonwealth Government in respect of goods purchased in Australia and which are, or which are intended to be, the property of the Telescope Board before going into use or consumption in Australia, and which are required in connection with the manufacture, construction, operation, use or maintenance of the telescope and its ancillary equipment.

(3) Goods which are owned by the Telescope Board at the time they are entered for home consumption or are purchased by the Telescope Board in Australia, or of which the Telescope Board acquires ownership before they go into use in Australia, shall remain its property and shall not be disposed of in Australia except under conditions acceptable to the Commonwealth Government.

(4) In accordance with the laws for the time being in force in Australia, the Commonwealth Government shall wherever possible facilitate the temporary admission into, or exit from, Australia free from Customs duties and other taxes, of goods or equipment which are required by a user of the telescope for the use in connection with the operation and use of the telescope.

(5) In accordance with the laws for the time being in force in the United Kingdom, the United Kingdom Government shall wherever possible facilitate the entry into, or exit from, the United Kingdom free from customs duties and other taxes, of goods which are required in connection with the manufacture, construction, operation and use or maintenance of the telescope and its ancillary equipment.

Article 21

In accordance with the laws for the time being in force in Australia, the Commonwealth Government shall, subject to current immigration policies and requirements, facilitate the admission into and exit from Australia of persons not normally resident in Australia employed or engaged as staff, consultants or advisers by the Telescope Board or by its contractors and of persons duly authorized to have access to or use of the telescope.

Article 22

The staff of the Telescope Board and its consultants, advisers and contractors and persons duly authorized to have access to or use of the telescope who are not normally resident in Australia but are required for the purposes of this Agreement to reside in Australia shall be permitted to import their personal and household effects on first arrival and to export those effects on departure free from import duties and taxes in accordance with the laws in force in Australia at the time of importation or exportation.

Article 23

This Agreement shall come into force on the date the Commonwealth Government notifies the United Kingdom Government that it has completed the processes necessary in Australia to give effect to the Agreement.

Article 24

(1) This Agreement shall continue in force for a period of 25 years unless previously terminated by the agreement of both Contracting Parties. After the expiration of 25 years either Party may terminate the Agreement by giving to the other 5 years notice of their intention.

(2) In the event that this Agreement is terminated the Contracting Parties shall agree on the manner in which the telescope and other property of the Telescope Board shall be dealt with. If such assets are sold, the net proceeds of such sale shall be divided equally between the two Parties.

(3) The terms and conditions of this Agreement may be varied or extended from time to time in such manner as may be agreed between the Contracting Parties.

IN WITNESS WHEREOF the undersigned, being duly authorized thereto by their respective Governments have signed the present Agreement.

DONE in duplicate at Canberra this twenty-fifth day of September, One thousand nine hundred and sixty-nine.

MALCOLM FRASER	C. H. JOHNSTON
FOR THE GOVERNMENT OF THE COMMONWEALTH OF AUSTRALIA	FOR THE GOVERNMENT OF THE UNITED KINGDOM OF GREAT BRITAIN AND NORTHERN IRELAND

NOTES

1. The *Anglo-Australian Telescope Agreement Act* 1970 (*a*) as shown in this reprint comprises Act No. 57, 1970 as amended by the other Acts specified in the following table:

Act	Number and year	Date of Assent	Date of commencement	Application, saving or transitional provisions
Anglo-Australian Telescope Agreement Act 1970	57, 1970	1 Sept 1970	22 Feb 1971 (*see Gazette* 1971, p. 1358)	
Anglo-Australian Telescope Agreement Act 1971	51, 1971	25 May 1971	1 Sept 1971 (*see* s. 2 and *Gazette* 1971, p. 5496)	
Statute Law Revision Act 1973	216, 1973	19 Dec 1973	31 Dec 1973	Ss. 9 (1) and 10
Administrative Changes (Consequential Provisions) Act 1978	36, 1978	12 June 1978	12 June 1978	S. 8

(*a*) This citation is provided for by the *Amendments Incorporation Act* 1905 and the *Acts Citation Act* 1976.

2. S. 16—Now cited as the *Compensation (Commonwealth Government Employees) Act* 1971.

Printed by Authority by the Commonwealth Government Printer

Appendix 2   Submission to the two Governments for management of the AAT

Proposed administrative arrangements for the AAT submitted to the two Governments by the Royal Society of London and the Australian Academy of Science in March 1965.

The requirement is for a form of administrative organization to meet the following conditions:

1. The capital monies for the construction and installation of the telescope will be provided jointly by an agreement between the Governments of Australia and Great Britain.

2. The annual running charges for staff, incidental equipment, administrative and maintenance charges, travelling, and the like, will also be provided within the agreement between these two Governments.

3. A permanent staff consisting of a director, scientific and other personnel is to be employed.

4. There must also be an arrangement for visiting scientific workers.

5. There will need to be some arrangement to allow the donor Governments to take part in the policy administration of the project.

6. For the day-to-day administration, the director would need to operate under Australian law, according to Australian administrative practices (i.e. in respect of salaries, superannuation, and the like) and having regard to UK practices.

It is proposed that an Australian–United Kingdom Astronomy Research Institute be set up by a convention between the two Governments, subsequently ratified by legislation, on a joint equal basis. Such an Institute would be made the responsibility of a board consisting of some 6–8 persons representing the two Governments equally.

Under such an arrangement the board of the Institute would have the power to undertake astronomical research, to appoint staff and manage the affairs of the Institute (including the construction of the telescope). The Director would be responsible to the Board for the day-to-day running of the Institute.

It is proposed that the observing time on the telescope should be at the disposal of the Institute staff, visiting Australian astronomers and visiting UK astronomers in equal proportions. A Research Committee, chaired by the Director, would be empowered by the Board to decide the relative merits of observing programmes proposed by prospective visiting astronomers and to allot time accordingly within the above limits. The Director would be solely responsible for the observing time allotted to the Institute staff.

Appendix 3　Membership of the JPC and the AAT Board

This chart shows the membership of the Joint Policy Committee until 1972, and of the Anglo-Australian Telescope Board from 1972 onwards. The vertical lines before the name of Street represent a short gap in membership following Eggen's resignation in 1973.

	Joint Policy Committee					Anglo-Australian Telescope Board				
	1967	1968	1969	1970	1971	1972	1973	1974	1975	1976
Members 3 UK	Hosie									
	Woolley									
	Bondi	Hoyle				Burbidge	Robins	Reddish	Massey	
3 Aust	Jones						Ennor			
	Bowen						Wild			
	Eggen						=	‖ Street		
Director								Wampler		Morton
Secretary	Cunliffe							Kazs		

	1977	1978	1979	1980	1981	1982	1983	1984	1985	1986
Members 3 UK	Robins	Atkinson								
	Reddish		Graham-Smith							
	Massey					Longair	Wilson			
3 Aust	Ennor	Farrands				Tegart				
	Wild					Frater				
	Street	Carver								
Director	Morton									Cannon
Secretary	Bryant			Franklin			Proust			

Appendix 3 *(continued)*

Anglo-Australian Telescope Board									
1987	1988	1989	1990	1991	1992	1993	1994	1995	1996
Members									
3 UK Atkinson	Egginton								
Longair	Graham-Smith								
Wilson		Boksenberg							
Tegart									
3 Aust Frater									
Carver		Cram							
Director Cannon									
Secretary Proust		Harrison							

Appendix 4   Biographical sketches

These notes are confined to people who were closely concerned with the telescope during its construction phase, either through the Joint Policy Committee, the Board or the Project Office. A number of other people are mentioned in the text.

E.G. Bowen, CBE FRS (1911–)   is a Welshman who joined the original British radar team in 1935, was in the USA for much of the war, and in 1946 went to the Australian CSIRO Division of Radiophysics where he remained. He was Chief of the Division during the exciting years when, together with the groups at the Universities of Cambridge and Manchester, it virtually created the new science of radiophysics. Five members of the Division, including himself, were later elected to the Royal Society. He played a central part in the funding and construction of the 210-foot Parkes radio telescope, a highly innovative and successful instrument. He was Chairman of the JPC and the AAT Board for the majority of the time between 1969 and 1973, and it was he more than anybody else who attracted the staff and created the conditions under which the AAT could be built to such a degree of excellence.

Professor E.M. Burbidge, FRS (1922–)   was born in England and studied at University College London before going to the USA in 1932 with her equally renowned husband, Geoffrey Burbidge. In 1972 she succeeded Woolley as Director of the RGO, and also took his place on the AAT Board during the years 1972–74 when the ANU dispute came to a head. She was a leading observational astrophysicist, and particularly at that time her wide experience of working at major observatories with the most advanced equipment available were of great value to the Board. At that time she had worked closely with Fred Hoyle and Joe Wampler.

D.W. Cunliffe, AM (1927–)   is an Australian who was working with the CSIRO Division of Mechanical Engineering in Melbourne when early in 1968 E.G. Bowen invited him to join the Project Office as Executive Officer. Later he became the Board's first Secretary, and in due course Executive Officer of the Anglo-Australian Observatory, giving him an unbroken association with the AAT for over 30 years. His sound judgment, even-handed and good-natured approach, engineering background and general competence were major factors in the smooth running of the Project Office. His initial and continuing contribution to the AAT won him an Australia Medal in 1990.

O.J. Eggen (1917–)   is an astronomer of the old school whose life and life-style have been determined by astronomy. A compulsive observer, he must have spent more time at the telescope than almost any other astronomer this century. He was born and educated in the USA and despite extended periods in England, Australia and Chile he remains invincibly American. In England his position at the RGO brought him into close association with Woolley. His strong personality, forthright style and refusal to compromise with what he saw as the best interests of his subject earned him powerful friends and powerful enemies. He served five years with the Office of Strategic Services, the precursor of the American Central Intelligence Agency, for much of that period in wartime Europe.

Sir Hugh Ennor, CBE FAA (1912–1977) was a biochemist from Melbourne. He was appointed to the first Foundation Chair in the Australian National University, and in 1957 became the first Director of its John Curtin School for Medical Research. One of the leading Australian scientists of his day, he was also a highly effective organiser and advocate, and as such found himself increasingly drawn into the administrative side of science. He finally relinquished his chair to become head of the newly created Department of Science, and joined the AAT Board in 1973. He became well-known for his efforts in the later stages of the ANU–AAT dispute, and was also prominent in the negotiations which saw Epping established as the site for the AAO laboratory.

Professor S.C.B. Gascoigne, ARAS FAA (1915–   )   was born in New Zealand, studied astronomical optics at the University of Bristol, and worked several years with the war-time optical munitions project at Mount Stromlo. After the war he set out to be a research astronomer with optics as only a side interest, but found himself increasingly involved with the re-equipping of Mount Stromlo Observatory as a stellar observatory, and he played an active part in refurbishing or commissioning a number of large telescopes. He was appointed Australian adviser to the Board, where he found his and Redman's interests dovetailed neatly. Redman's special fields were the dome and building and all other aspects of seeing, and the mechanical aspects of the mounting, Gascoigne's were optical design and the telescope control system. Both were deeply concerned with the manufacture and testing of the optics.

W.A. Goodsell (1920–1987)   entered the British Ministry of Works in 1949, and worked his way up through the engineering grades, gaining experience in a wide variety of projects, many related to aviation. A major one was a large low-velocity wind tunnel. In 1970 he was appointed Project Manager of the AAT for three years. When his term ended in 1973 he returned to England to become Project Manager for the William Herschel Telescope. This telescope was successfully completed in 1987. Besides his unusual breadth of engineering experience, Goodsell was an expert on project management and contract law. John Pope said 'In all the meetings I attended with him I never once saw a contractor get the better of Bill Goodsell'.

J.F. Hosie, CBE (1913–   )   is a Scotsman who studied at the Universities of Glasgow and Cambridge. He joined the Indian Civil Service before the war, and worked as a District Officer and then in GCHQ India until Indian Independence in 1947. Subsequently a career in the British Administrative Civil Service led to a post as a Director in the Science Research Council. Here his skills were deployed in formulating and controlling administrative and financial aspects of British participation in international scientific projects, particularly in space science and astronomy. As a founder member of the AAT Board, he and his Australian colleague K.N. Jones forged the administrative and financial framework embodied in the AAT Agreement. He was a forceful and determined upholder of British interests, and was dedicated to the success of the AAT.

Professor Sir Fred Hoyle, FRS (1915–   ) is a Cambridge product and one of the best known scientists of his time. Some of his contributions, notably in stellar evolution, the nucleosynthesis of the elements, and theoretical cosmology, have long since been an integral part of modern astrophysics. As Board Chairman he earned much credit for the skill with which he steered the Project through some deeply troubled waters, and for the way he could come to grips with difficult technical problems which must initially have been quite unfamiliar to him. He has written a typically frank, informal account of his years on the Board, and some successful science fiction novels, in one of which Siding Spring is portrayed as 'Wombat Spring'.

M.H. Jeffery (1926–1969) was a structural engineer by training, a member of the British firm of consulting engineers, Freeman Fox and Partners. Bowen first met him in 1954 when he was working with Freeman Fox on the design of the 210-foot Parkes radio telescope. He accompanied Bowen on his tour of the principal tenderers for the telescope, then went to Parkes as Resident Engineer for the erection and commissioning phase. He became widely known, and Bowen considered he could well have succeeded Bruce Rule as doyen of the world's telescope engineers. As the first AAT Project Manager (1968–69) he was a fine leader and a first-class engineer, and his death was a loss to the profession as well as to the Project.

K.N. Jones, CBE (1924–   ) had a distinguished career in the Australian Public Service. In April 1967 when attached to the Department of Education and Science, he travelled to London with O.J. Eggen to play a leading part in the meeting with the SRC which resulted in the agreement to build the telescope. Together with J.F. Hosie, he was an original member of the JPC and the Board, and had much to do with drafting of the AAT Agreement. He relinquished his membership of the Board when his Department's responsibilities were reorganised in 1973. He was a steadfast, level-headed man who represented his country most capably, sometimes in extremely difficult circumstances.

Professor Sir Harrie Massey, FRS (1908–1983)   An Australian by birth, he was a convinced Anglo-Australian and a great internationalist. His outstanding academic abilities combined with vision, foresight and determination, and an enviable imperturbability enabled him to move easily in the corridors of power in Britain and Australia. At the same time he would inspire scientists with his enthusiasm and persistence in the face of non-scientific obstacles. His career from the University of Melbourne via Cambridge and Belfast to the Quain Professorship of Physics at University College London is studded with distinctions. Both informally, and later as a member of the Board and ultimately its Chairman, he made good use of his contacts in both countries. He argued cogently and successfully for adequate funding for the project, smoothing away difficulties and always ensuring that the AAT remained in the highest class of observatories.

H.C. Minnett (1917–   )   began his career in the CSIRO Radiophysics Laboratory working on wartime radar, and finished it as Chief of the Division. In between he worked on radio astronomy, antennas, and aircraft navigation and landing systems. He was associated with the Parkes telescope from its inception, spending several years in London as CSIRO representative while the engineering studies were in progress. He was a central figure in the design of the AAT drive and control systems. He was Project Manager 1969–70.

J.D. Pope (1924–   )   was trained initially at C.A. Parsons and Co., the parent company of Sir Howard Grubb, Parsons and Company. He joined the RGO in 1944, and worked in the Time Department until Woolley's arrival at Herstmonceux in 1955. Then he was given responsibility for all engineering work at RGO, with his first task the reassembly of all the telescopes which had been moved there from Greenwich. Like Wehner's, his association with the AAT goes back to the early 1960s, and he was concerned with almost every aspect of it. In 1971 Pope left the AAT (though maintaining active connections with it) to begin work on the William Herschel Telescope. He saw it through to completion at Grubb Parsons' works before he retired in 1984.

Professor V.C. Reddish, OBE FRSE (1926–   )   is an astronomer whose career from University College London, through the Universities of Edinburgh and Manchester, led to his eventual appointment as Regius Professor in the University of Edinburgh, Director of the Royal Observatory Edinburgh, and Astronomer Royal for Scotland. He was responsible for the Schmidt Telescope Project at Siding Spring, creating a new design to the same scale as the Palomar design. As a consultant he was a major contributor to the initial instrumentation programme for the AAT. He also has made outstanding contributions to British telescopes in Hawaii and elsewhere. As a Board member he was able to bring his experience to bear on many of the difficult practical problems which faced the Board as the telescope reached the operational stage.

Professor R.O. Redman, FRS (1906–1975)   was Professor and Director of the Observatories at Cambridge. He had long been recognised as the astronomer who when it came to telescope practicalities was without a peer in Britain, and was the obvious choice as British astronomical adviser to the Board. He acquired his experience on the 72-inch telescope at the Dominion Astrophysical Observatory in British Columbia, on the 74-inch at Pretoria, for which he designed a much used spectrograph, and from the post-war rehabilitation of his own observatory. He had a natural engineering sense which was highly respected in the Project Office, and combined this with a sure touch with people, especially as a pourer of oil (but never too much) on troubled waters.

M.O. Robins, CBE (1918–   )   After studying physics at Oxford University, he was involved in wartime work on aircraft armaments, leading after the war to pioneering work on guided rockets at Farnborough. Subsequently he joined Sir Harrie Massey at University College London, and as the British Project Officer for the first joint US–UK scientific satellite programme, he played a significant part in the development of the British space science programme. He followed Hosie as a Director in the Science Research Council, again concerned with astronomy, and this led to his involvement with the AAT as a Board member. His experience in both science and administration was particularly appropriate at a time when the problems of locating the laboratory and of building up the initial staff were causing many difficulties.

Professor R. Street, AO FAA (1920–   )   was born in England and came to Melbourne in 1960 as Professor of Physics in Monash University. From 1974 to 1978 he was Director of the ANU Research School of Physical Sciences, one component of which was Mount Stromlo Observatory. He joined the Board in 1974, to become deeply concerned with the choice of the Epping site and setting up the operational framework of the new telescope. At the same time, he did much to restore the somewhat tarnished image of the ANU in the Australian scientific world.

H.P. Wehner (1924–   )   was born in Germany and worked initially at the Göttingen Observatory, where with Dr H. Haffner he designed and built an instrument called an iris photometer which was widely used at many observatories. After studying in Munich he emigrated to Australia in 1952, to a position at Mount Stromlo Observatory, where like his opposite number John Pope at the RGO he became immersed in optical telescopes for many years. His long association with the AAT began in the early 1960s. He carried the major responsibility for the dome and building, though he was concerned with most aspects of the telescope, and was Project Manager 1973–75. More recently he was associated closely with the construction of Mount Stromlo's 2.3-metre alt-azimuth telescope.

J.P. Wild, AC CBE FTS FAA FRS (1923–) is one of the most distinguished and influential Australian scientists of his generation. Born in Yorkshire, he studied at Cambridge, and from 1943 to 1947 served with the Royal Navy as a radar officer. In 1947 he joined the CSIRO Radiophysics Division in Sydney, where he quickly made his name as a leading pioneer of solar radio astronomy. He has long been recognised as one of the founding fathers of Australian radio astronomy. A high point in his career was the commissioning of the solar radio heliograph at Narrabri, New South Wales, at what is now the Paul Wild Observatory. He succeeded E.G. Bowen as Board member in 1973, and became Board Chairman in 1975.

Professor Sir Richard Woolley, OBE FRS (1906–1986) was a brilliant, many-talented man of great presence and style, equally at home in scientific, academic and (when necessary) governmental circles. Of joint British and South African descent, he was a product of the classical British astronomical schools of Cambridge and Greenwich. From the age of 33 he was Director in succession of the Mount Stromlo, Royal Greenwich and South African Observatories, leaving an enduring mark on each. The concept of the AAT was essentially his, and his early advocacy for it was of great importance because it kept the project alive during a critical period when it might well have perished. He was one of the original members of the JPC, and on the AAT Board until 1972.

Appendix 5 Staff of the Project Office

	Position	Term
Ronald Adams	Engineer	until 1972
Peter Anyon	Engineer	1972 to 1975
David Barrott	Draughtsman	1968 to 1973
Maston Beard	Head of computer group	1971 to 1974
Graham Bothwell	Computer engineer	1968; then went to AAO
Cheryl Christiansen	Secretary	1972 to 1975
Douglas Cunliffe	Executive Officer	1968; then went to AAO
S.C. (Ben) Gascoigne	Adviser on astronomy Commissioning astronomer	pre-1967 to 1973 1974 to 1975
Peter Gillingham	Engineer	1971; then went to AAO
William Goodsell	Third Project Manager	1970 to 1973
Ken Hall	Structural engineer	1969 to 1972
Peter Hewitt	Instrument designer	1970 to 1974
Michael Jeffery	First Project Manager	1968 to 1969
Henry Kobler	Electronics engineer	1973 to 1975
Harry Minnett	Consultant engineer Second Project Manager	from 1967 1969 to 1970

Adrian Mortimer	Computer programmer	1973
Adrian Mortimer	Computer programmer	1973
David Murray	Draughtsman	1972 to 1974
Chris Nixon	Personnel officer	1972
John Pope	Senior engineer	pre-1967 to 1971
John Rock	Draughtsman	1971; then went to AAO
Jack Rothwell	Electrical engineer	1968 to 1975
John Straede	Computer programmer	1973; then went to AAO
Heather Thomson	Clerical assistant	1970 to 1975
Patrick Wallace	Computer programmer	1973; then went to AAO
Tom Wallace	Computer programmer	1971 to 1973
Herman Wehner	Senior engineer Final Project Manager	pre-1967 to 1973 1973 to 1975

Appendix 6 Statistical facts of the AAT

Altitude of site	1165 metres
Latitude	31°16′37″S
Longitude	149°3′58″E
Overall diameter of primary mirror	3.94 metres
Working diameter of primary mirror	3.89 metres
Diameter of central hole	1.070 metres
Thickness of primary mirror at outside edge	0.63 metres
Thickness of primary mirror at central hole	0.56 metres
Focal length of primary mirror	12.70 metres
Mass of primary mirror	16.19 tonnes
Mass of primary mirror plus cell and support system	43.70 tonnes
Mass of telescope tube assembly	110 tonnes
Mass of polar axis assembly (including tube)	258 tonnes
Mass of base frame assembly	68 tonnes
Total mass of telescope	326 tonnes
Diameter of horseshoe mounting	12.19 metres
Diameter of dome	36.54 metres
Weight of dome	approx. 570 tonnes
Maximum speed of rotation of dome	72°per minute
Height of top of dome above ground level	50.29 metres

Appendix 7   Major contracts

Dome and building

Macdonald, Wagner and Priddle Sydney, Australia	Consulting engineers for the design of the dome and building
Leighton Contractors Limited Sydney, Australia	Contractors for the main building
Evans Deakin Limited Brisbane, Australia	Sub-contractors for construction of the dome
Hodgson and Lee Pty Limited Sydney, Australia	Contractors for electrical cabling of the building
W.E. Bassett and Partners Melbourne, Australia	Consulting engineers for the air-conditioning system
Athertons (NSW) Pty Limited Melbourne, Australia	Contractors for the air-conditioning
Edwards High Vacuum Limited Crawley, England	Manufacturer of the aluminising plant

The telescope

Freeman Fox and Partners London, England	Consulting engineers for the telescope mounting
Dilworth, Secord and Meagher Toronto, Canada	Consulting engineers for the telescope tube and declination axis assembly
Maag Gear-wheel Company Limited Zurich, Switzerland	Machining of the main drive gears
Ransome, Hoffmann and Pollard Bearings Limited, Chelmsford, England	Manufacturer of the declination bearings

GEC–AEI (Electronics) Limited   Designer of the mechanical and electrical
Leicester, England   drive and control systems

E.A. Sweo and Associates   Designer of the mechanical and electrical
San Francisco, USA   drive and control systems

Mitsubishi Electric Corporation   Manufacturer of the mounting, the mech-
Japan   anical drive system, and the electronic
                                       system for the drive and control

GEC–Elliott Process Automation   Designer of the computer system
Limited, Leicester, England

Datronics Systems Pty Limited   Supplier of the Interdata computer system
Sydney, Australia

Optics
Owens–Illinois Inc   Manufacturer of the mirror blanks
Toledo, Ohio, USA

Sir Howard Grubb, Parsons   Figuring of the mirrors, and design
and Company Limited   and manufacture of the telescope tube
Newcastle–upon–Tyne, England

W.E. James   Manufacturer of the prime focus correctors
Melbourne, Australia

Ancillary equipment
Bendigo Ordnance Factory   Manufacturer of the prime focus camera
Bendigo, Australia   and guider, coudé No 5 drive

Maribyrnong Ordnance Factory   Manufacturer of the finder telescope
Melbourne, Australia

RAAF Academy at the University   Chopping secondary mirror for infrared
of Melbourne, Australia   observations

SIRA Institute   Manufacturer of the Cassegrain acquisition
London, England   and guidance system, and autoguider
                                       electronics

Quantex Corporation
California, USA

Image intensifier integrating the closed
circuit television system

Principal instrumentation
Boller and Chivens Division of
Perkin–Elmer Corporation
Los Angeles, USA

Fast Cassegrain spectrograph

Digital Equipment Corporation, USA

VAX 11/780 and 3500 computer systems

Kapteyn Sterrenwacht, Groningen
The Netherlands

Imaging Fabry–Perot interferometer
(Taurus)

Mount Stromlo and Siding Spring
Observatories, Australia

Photometers and spectrometer and
the Fabry–Perot Infrared Grating
Spectrometer (FIGS)

Royal Greenwich Observatory
England

Cassegrain spectrograph, and the first
CCD system

University College London
England

Image Photon Counting System (IPCS),
and coudé echelle spectrograph (UCLES)

University of Durham, England

automatic fibre positioning system (Autofib)

Epping Laboratory
Moore and Cashell

Architects for the Epping Laboratory

Appendix 8 Speeches at the AAT inauguration

Inaugural address by H.R.H. The Prince Charles

Not being in any way an astronomer myself I was not properly aware of the great importance of this new Anglo-Australian Telescope here at Siding Spring when I was asked to come out to Australia and officially declare it open. But you may be glad to hear that I have been acquainting myself with the significance of Australian research in the field of astronomy and with the history of the close co-operation between Australian and British astronomers since the earliest days. I have become totally confused by stellars, pulsars and quasars, by the difference between galactic astronomy and interstellar atomic and molecular spectroscopy – by such fascinating sounding things as modulated synchroton radiation. My interest has been roused and I am only sad I cannot stay here longer and receive some lessons in basic astronomy from the Director. I might also find a few days here most enlightening and educative from the point of view of my naval experience. The nearest I have ever come to astronomy is in astro-navigation and I cannot claim *that* to be my favourite or most skilful occupation. As I am sure you will realise, if you make one mistake with the operation of a sextant or a small miscalculation when you come to work out your sights then your resultant position on the plotting sheet can be out by as much as a whole degree of latitude. The ensuing embarrassment is intense if this leads to missing, for instance, the island of Bermuda or hitting Canada rather than the United States of America! So you can perhaps see that a period of instruction here could be most beneficial to one of the Royal Navy's worst navigators.

The history of astronomical observation in Australia is a long and important one – and one that has contributed greatly to the present level of understanding of our own galaxy. It began when the Governor of New South Wales, Sir Thomas Brisbane, arrived in Australia in 1821 bringing with him the necessary instruments for an observatory together with two assistants. It was probably Brisbane's passion for astronomy and the fact that no observations of stars had been made in the southern hemisphere since the work of Lacaille at the Cape of Good Hope in 1751–52, that induced him to apply for the post of Governor. When the government declined him assistance for an observatory, he purchased all the instruments himself and set up his observatory near Government House, Parramatta. I am delighted and relieved to see that the Governments of Australia and Britain are more enlightened now in the field of scientific research than the then British Government showed itself to be in 1821 – or perhaps Sir Thomas Brisbane should have been more persistent in his representations to British bureaucrats. Whatever the case, this whole development here at Siding Spring is a splendid example of the best of British and Australian expertise in the realm of astronomical observation – and with an American Director of such noteworthy scientific repute its future as a centre of international study must indeed be assured. I am sure too that its (dare I say it?) astronomical cost will be more than justified by the use that is made of it in the pursuit of further knowledge of our universe. For it is only by continued exploration into hitherto undreamt of realms of scientific research – particularly with regard to the structure of our galaxy – that we can face the rapidly increasing limitations of our own earthly existence – and overcome them to the advantage of everyone.

Realising what a tremendous feat of engineering this whole telescopic complex has been, I would like to congratulate, with admiration, all those who have worked so hard to set up the whole operation and keep it running, and who must be praying now that the instrument will work when required. In appreciation of the importance of this telescope in the southern hemisphere, it is particularly exciting for me to be able to declare this aperture open.

Speech by the Australian Prime Minister, E.G. Whitlam

There could be no more fitting symbol of the Anglo-Australian relationship than this telescope. It commemorates a spirit of scientific endeavour – and a history of astronomical research – that began with Cook's first voyage to the South Seas. The astronomers of the Royal Society in the late eighteenth century set out to observe the transit of Venus. Cook finished up charting the east coast of Australia. Astronomers – who in such matters, as in all things, take the long view – will doubtless continue to argue which of these phenomena was more important.

The fact remains that the colonisation of Australia by Europeans had its origins in an astronomical project. Trade followed the flag, but the flag followed the telescope. It may not be too romantic to suggest that the telescope on the *Endeavour* was as potent a symbol of Britain's imperial greatness as Nelson's telescope on the *Victory*. More than two centuries after Cook's voyage, British and Australian astronomers can now pursue their observations of the southern skies with this new Anglo-Australian Telescope – this brilliant example of advanced technology and precision engineering which we see before us.

I commend Professor Hoyle's view that 'a telescope is a good example of the things which our civilisation does well'. It is a good example of the things a true civilisation alone can do. Astronomy is one of the few scientific pursuits – certain branches of mathematics, certain branches of theoretical nuclear physics are others – which exist for no other purpose than the pure search for knowledge. No one doubts that astronomy has its practical uses. Governments are frequently reminded of them by astronomers themselves. Nevertheless, the true glory of astronomy lies in the value it places on knowledge for its own sake – knowledge of an absolute and fundamental kind. No civilisation can remain indifferent to the origins of the earth or its place in the cosmos.

Astronomy is rich in theories but somewhat deficient in practical results. In that respect it can be likened to economics. In its attempt to provide answers to the ultimate questions about man's environment and his place in reality, it can be likened to religion. It is no accident that the high priests of astronomy conduct their nocturnal rites in remote temples on inaccessible mountain tops. So I think there is an excellent case for Government support of astronomy, just as we give State aid to religious enterprises. We are not alone in this. The Americans, a pragmatic and hard-headed people, have provided vast public funds for a space program whose ultimate purpose is to add to man's knowledge of the cosmos and his place within it. I would not wish to justify the scale of that expenditure, but I approve wholeheartedly the purpose behind it.

Government support for astronomy has a long history. It can be traced from the ancient civilisations to the present day. From the Egyptians, the Persians, Copernicus, Galileo, Newton and Kepler, to the platform of the Australian Labor Party. Our policy on science pledges 'particular support to areas in which Australia can make a special contribution, including astronomy . . . '.

In the eighteenth and nineteenth centuries, astronomy was recognised for its importance to navigation and cartography. It was sufficiently important at the time of Federation for the founding fathers to make specific provision in the Australian Constitution for the Parliament to make laws with respect to astronomical observations. Two such laws have been made: in 1923 when Parliament established the Commonwealth Solar Observatory, and in 1970 when it passed legislation setting up the Agreement for this telescope at Siding Spring. It is appropriate that the study of the laws of the planets and stars should be mentioned in the Australian Constitution – a document as timeless and immutable as the heavenly bodies themselves.

The present Australian commitment to astronomy is extensive. The Australian National University operates major optical telescopes at Mount Stromlo and here at Siding Spring. Major radio telescopes are operated by the CSIRO and by Sydney University. The University of Tasmania has a growing involvement in astronomy, and there are State Government observatories in Sydney and Perth. The Australian National University astronomers have gained a world-wide reputation for their work on the structure and evolution of galaxies, stellar structure and the chemical composition of stars. Radio astronomers in CSIRO and Sydney University have made important contributions to quasar astronomy and to the study of the structure and evolution of the universe.

Groups working with the Parkes radio telescope have recently discovered many different species of organic molecules in the gas clouds of interstellar space. These discoveries may throw light on the processes that led to the existence of life on earth. Astronomers at Sydney University have performed work of fundamental significance in the measurement of the sizes of stellar objects. The budget last month provided a grant to enable a study of a new instrument to extend this work.

Astronomy for many years has enjoyed a wide measure of international co-operation both at the institutional and the individual level and Australian astronomy has benefited from this. Australia and Britain are sharing equally the capital cost of this telescope, amounting to $16 million, and the estimated annual operating cost of $1.3 million. Many of the astronomical facilities in Australia were built with funds contributed partly from abroad, and many of the astronomers working in Australia were born overseas. Some are here as visitors and some are permanent residents. They come from a variety of countries including Sri Lanka, India, Korea, the People's Republic of China, Israel, Poland, the United Kingdom, Germany, USA, New Zealand and Sweden. The Director of this telescope, Dr E.J. Wampler, is an American citizen and two of the three Australian members of the AAT Board, Dr Wild and Professor Street, were born in the United Kingdom. There are many Australian astronomers working overseas.

I am proud to be associated with the inauguration of this telescope. I am proud that my Government is contributing to its cost. Its beauty, its power, its precision reflect the highest credit on all who designed and built it. Those who use it, like politicians, will sit through the long reaches of the night. You will contribute to man's understanding of the earth and the universe. The questions you ask are of transcendental importance to our knowledge of ourselves and our surroundings. It may well be that the ultimate answers you seek will always elude you; that in the last resort the mysteries you attempt to unravel will remain in the philosopher's court. For all that, it is good to reflect that here in this building and in others like it, among you and your colleagues, scientific research in its purest, most peaceful and most disinterested form will go forward for the enrichment of our understanding and for the benefit of future generations.

Appendix 9 The world's largest telescopes

Name	Location	Aperture cm	ins	f/no	Mirror	Year	Mounting	Reference to Sky and Telescope	
1 Hale	California	508	200	3.3	pyrex	1948	horseshoe	(1951)	**10** 186
2 Lick	California	305	120	5.0	pyrex	1959	fork	(1956)	**16** 62
3 KPNO	Arizona	381	150	2.8	silica	1973	horseshoe	(1973)	**45** 10
4 AAT	New South Wales	389	153	3.26	Cervit	1975	horseshoe	(1975)	**50** 225
5 CTIO	Chile	400	158	2.8	Cervit	1976	horseshoe	(1976)	**51** 72
6 BTA	Caucasus	600	236	4.0	pyrex	1976	alt-azimuth	(1977)	**54** 356
7 ESO	Chile	357	141	3.0	silica	1977	horseshoe	(1977)	**53** 97
8 MMT	Arizona	450	177	2.7	multiple	1979	alt-azimuth	(1976)	**52** 14
9 UKIRT	Hawaii	380	150	2.5	Cervit	1979	yoke	(1978)	**56** 22
10 CFHT	Hawaii	360	142	3.8	Cervit	1979	horseshoe	(1977)	**53** 254
11 NASA	Hawaii	300	118	2.5	Cervit	1979	yoke	(1980)	**60** 462
12 MPI	Spain	350	138	3.5	Zerodur	1984	horseshoe	(1986)	**72** 25
13 WHT	La Palma	420	165	2.5	Cervit	1987	alt-azimuth	(1987)	**74** 458
14 NTT	Chile	358	141	2.2	Zerodur	1989	alt-azimuth	(1988)	**75** 464
15 ARC	New Mexico	350	138	1.75	Angel	1989	alt-azimuth	(1988)	**76** 126
16 Keck	Hawaii	1000	400	1.75	mosaic	1991	alt-azimuth		

Year refers to the year of commissioning.

References to *Sky and Telescope* are year, volume, page; they follow Chassen J. and Sperling N. (1981). *Sky and Telescope* **61**, 303

BTA : the Zelenchuk telescope of the USSR Academy of Sciences

MMT : the Multi Mirror Telescope of the Smithsonian Institute and University of Arizona

UKIRT : the United Kingdom Infrared Telescope operated by the Royal Observatory, Edinburgh, on Mauna Kea

CFHT : the Canada–France–Hawaii Telescope on Mauna Kea

NASA : an infrared telescope operated by NASA on Mauna Kea

MPI : the telescope of West Germany's Max–Planck–Institut für Astronomie

WHT : the British William Herschel Telescope in the Canary Islands

NTT : the New Technology Telescope of the European Southern Observatory

ARC : the Astrophysical Research Consortium of five American universities

A voluminous literature has grown up about large telescopes. Further reading on the telescopes as numbered in the preceding table may be found:

1, 2 *Telescopes* (1960). G.P. Kuiper & B. Middlehurst (ed.) University of Chicago Press

   6 Ioannisiami B.K. *et al. Instrumentation for Astronomy with Large Optical Telescopes* (1981). C. Humphries (ed.) IAU Colloquium No 67, Reidel p.3

   7 Richter W. *et al. ESO Technical Reports* (1974). No 1 *et seq.*

   8 *Telescopes for the 1980s* (1981). G. Burbidge & A. Hewitt (ed.) Annual Reviews Inc.
      Advanced Technology Optical Telescopes III (1986). L.D. Barr (ed.) *Proc. SPIE* **628**, section 1

   9 Humphries C.M. *Optical and Infrared Telescopes for the 1990s* (1980). A. Hewitt (ed.) p.487

  10 Bely P. & Lelievre G. *Instrumentation for Astronomy with Large Optical Telescopes* note 6, p.21

  13 Murdin P.G. *Instrumentation for Astronomy with Large Optical Telescopes* note 6, p.49
      Boksenberg A. (1985). *Vistas in Astronomy* **28**, 531

  14 Tarenghi M. *Advanced Technology Optical Telescopes III* note 8, p.213

  15 *Advanced Technology Optical Telescopes III* note 8, pp.369–402

  16 Gustafson J.R. & Sargent W. (1988). *Mercury* **2**, 43

Bibliography

Allen C.W. (1973). *Astrophysical Quantities,* The Athlone Press, London

Aller L.H. (1963). *Science,* **139**, 21

Anglo-Australian Observatory (1976). Observer's Guide to the AAT

Anglo-Australian Telescope, Annual Reports 1970–71 to 1987–88

Australian Academy of Science (1980). *The First Twenty-five Years,* F. Fenner & A.L. Rees (ed.) The Anglo-Australian Telescope Project 47

Babcock H.W. (1978). Proc. ESO Conference on Optical Telescopes of the Future, F. Pacini *et al.* (ed.)

Bahner K. (1967). *Handbuch der Physik,* Vol. XXIX, Springer–Verlag

Bartholomew J. (1953). *Advanced Atlas of Modern Geography*

Bertin B. (1971). Proc. ESO/Cern Conference on Large Telescope Design, R.M. West (ed.) 395

Bok B.J. (Jan 1963). The Future of Astronomy in Australia *The Australian Journal of Science,* 281

Bowen E.G. (1981). The Pre-history of the Parkes 64-m Telescope *Proc. Astron. Soc. Aust.* **4**, 267

Bowen E.G. & Minnett H.C. (Feb 1963). *Proc. IRE (Aust),* **24**, 98

Bowen I.S. (1964). Telescopes *Astron. J.* **69**, 816

Bowen I.S. (1967). Astronomical Optics *Annu. Rev. Astron. Astrophys.* **5**, 45

Bowen I.S. (1967). Future Tools of the Astronomer *Qrt. J. R. Astr. Soc.* **8**, 9

Brown D.S. (1973). *Observatory,* **93**, 208

Burbidge G. (1972). *Nature* (London), **239**, 118

Burbidge G. & Barr L.D. (ed.) (1982). Advanced Technology Optical Telescopes, Proc. SPIE Conference No. 332

Burbidge G. & Hewitt A. (ed.) (1980). *Telescopes for the 1980s,* Annual Reviews, Palo Alto

Carroll J. (1972). *Qrt. J. R. Astr. Soc.* **13**, 132

Clerke A. (1893). *A Popular History of Astronomy during the Nineteenth Century,* Black, London

Cockburn S. & Ellyard D. (1981). *Oliphant: the Life and Times of Sir Mark Oliphant,* Axiom Books

Crawford D.L. (ed.) (1966). The Construction of Large Telescopes IAU Symp. No. 27, Academic Press

Crawford D.L. (1965). *Sky and Telescope,* **29**, 354

Danjon A. & Couder A. (1935). *Lunettes et Télescopes,* Editions de la Revue d'Optique theoretique et instrumentale, Paris

Di Cicco D. (1986). *Sky and Telescope,* **71**, 347

Dreyer J.L. & Turner H.H. (ed.) (1923). *History of the Royal Astronomical Society 1820–1920,* RAS, London

Evans D.S. & Mulholland J.D. (1986). *Big and Bright,* University of Texas, Austin

Fern D. (Dec 1975). *Metal Construction,* 600

Gascoigne S.C.B. (1963). Towards a Southern Commonwealth Observatory *Nature* (London), **197,** 1240

Gascoigne S.C.B. (1968). *Records of the Australian Academy of Science,* **1,** 65

Gascoigne S.C.B. (1968). Recent Advances in Astronomical Optics *Qrt. J. R. Astr. Soc.* **9,** 98

Gascoigne S.C.B. (1970). *J. Phys. E.* **3,** 167

Gascoigne S.C.B. (1984). The Woolley Era *Proc. Astron. Soc. Aust.* **5,** 597

Gascoigne S.C.B. (1988). *Australian Science in the Making,* R.W. Home (ed.) Cambridge University Press

Gillingham P.R. (1983). Advanced Technology Optical Telescopes II *Proc. SPIE* **444,** 165

Gillingham P.R. (1984). Proc. IAU Colloq. No. 79, Very Large Telescopes, their Instrumentation and Programs, M-H. Ulrich *et al.* (ed.) 415

Gillingham P.R. (1989). Lessons for New Large Telescopes from the AAT, *Japanese National Large Telescope and related engineering development,* T. Kogure & A.T. Tokunaga (ed.) 281

Gingerich O. (1976). *Vistas in Astronomy,* **20,** 2

Gingerich O. (ed.) (1984). *Astrophysical and Twentieth Century Astronomy to 1950,* in The General History of Astronomy Vol. 4, Cambridge University Press

Goodsell W.A. The History of the Anglo-Australian Telescope, in the archives of the Royal Greenwich Observatory, **47,** 12

Graham-Smith F. & Dudley J. (1982). *J. Hist. Astron.* **13,** 1

Hansard (10 April 1963). Commonwealth of Australia Parliamentary Debates, House of Representatives 579

Hansard (9 April 1963). Commonwealth of Australia Parliamentary Debates, Senate 349

Hardy G.H. (1937). *A Course in Pure Mathematics,* Cambridge University Press

Henbest N. & Marten M. (1983). *The New Astronomy,* Cambridge University Press

Hewitt A. (ed.) (1980). Proc. KPNO Conference on Optical and Infrared Telescopes for the 1990s

Hill J.M. (1988). *The History of Multi-object Fibre Spectroscopy, Astron. Soc. Pac. Conf. Series* **3,** 77

Hoyle F. (1982). *The Anglo-Australian Telescope,* University College Cardiff Press

Humphries C.M. (ed.) (1981). Proc. IAU Colloq. No. 67, Instrumentation for Astronomy with Large Optical Telescopes

King H.C. (1955). *The History of the Telescope,* Dover, New York

Kobler H. & Wallace P.T. (1976). *Publ. Astron. Soc. Pac.* **88,** 80

Kuiper G.P. & Middlehurst B. (ed.) (1960). *Telescopes,* University of Chicago Press

Lovell A.C.B. (1987). *Qrt. J. R. Astr. Soc.* **28**, 1

McCrea W.H. (1988). Richard van der Riet Woolley *Biog. Mem. Fel. R. Soc.* **34**

McCrea W.H. (1989). *Historical Records of Australian Academy of Science,* **7**, 315

Meinel A.B. (1969). *Applied Optics and Optical Engineering* – V, R. Kingslake (ed.), Academic Press, 136

Minnett H.C (1971). *Proc. Astron. Soc. Aust.* **2**, 2

Minnett H.C. (1962). *Sky and Telescope,* **24**, 184

Observatory (1934). **57**, 250

Observatory (1971). **91**, 89

Odgers G.J. & Wright K.O. (1968). *Journal RAS Canada,* **62**, 392

Pacini F. *et al.* (ed.) (1978). Proc. ESO Conference on Optical Telescopes of the Future

Pannekoek A. (1961). *A History of Astronomy,* Allen & Unwin

Plaskett H.H. (1946). Presidential Address *Mon. Not. R. Astr. Soc.* **106**, 80

Pope J.D. (1971). Optical Performance Criteria for Telescope Tube Design, Proc. ESO/CERN Conference on Large Telescope Design, R.M. West (ed.) 299

Preston R. (26 Oct 1987). *The New Yorker,* 81

Quick J. & Garran R.R. (1976). *The Annotated Constitution of the Australian Commonwealth,* Legal Books, Sydney

Redman R.O. (1960). Presidential Address *Qrt. J. R. Astr. Soc.* **1**, 10

Redman R.O. (1971). The Anglo-Australian Telescope *New Scientist* , **50**, 313

Robinson L.B. (1975). Online Computers for Telescope Control *Annu. Rev. Astron. Astrophys.* **13**, 165

Ross F.E. (1935). *Astrophys. J.* **81**, 156

Rothwell J. (1971). Proc. ESO/Cern Conference on Large Telescope Design, R.M. West (ed.) 445

Science and Engineering Research Council, Annual Reports

Straede J.O. & Wallace P.T. (1976). *Publ. Astron. Soc. Pac.* **88**, 792

Tayler R.J. (ed.) (1987). *History of the Royal Astronomical Society 1920–1980,* Vol. 2, Blackwell

Trumbo D. (1966). *The Construction of Large Telescopes,* D.L. Crawford (ed.) Academic Press, 131

Wallace P.T. (1978). Proc. ESO Conference on Optical Telescopes of the Future, F. Pacini *et al.* (ed.) 123

Wallace P.T. (1975). Proc. MIT Conference on Telescope Automation, 284

Warren S.J. *et al.* (1987). *Nature* (London), **330**, 453

West R.M. (ed.) (1971). Proc. ESO/CERN Conference on Large Telescope Design, CERN, Geneva

Whelan J.A. (1976). A Night at the Anglo-Australian Telescope *Qrt. J. R. Astr. Soc.* **17**, 306

Wood H. (1983). Sydney Observatory 1858–1983 *Proc. Astron. Soc. Aust.* **5**, 273

Wood H. (1974). Astronomical site prospects in New South Wales, Sydney Observa-

tory Papers No. 70
Woolley R.v.d.R. (1962). *Qrt. J. R. Astr. Soc.* **3**, 249
Wynne C.G. (1965). *Applied Optics*, **4**, 1185
Wynne C.G. (1972). *Progress in Optics*, **10**, 139

Index of names

Index of subjects